Military Aircraft
Pilot Reports

Other books by Aviation Week & Space Technology Magazine

Business & General Aviation Aircraft Pilot Reports
Commercial & Regional Transport Aircraft Pilot Reports
Helicopter Pilot Reports

Military Aircraft Pilot Reports

Aviation Week & Space Technology Magazine

McGraw-Hill

New York San Francisco Washington, D.C. Auckland Bogotá
Caracas Lisbon London Madrid Mexico City Milan
Montreal New Delhi San Juan Singapore
Sydney Tokyo Toronto

McGraw-Hill

A Division of The *McGraw-Hill* Companies

Library of Congress Cataloging-in-Publication Data
Military aircraft pilot reports / by Aviation week & space technology
   magazine.
      p. cm.
   Includes index.
   ISBN 0-07-003089-8 (P)
   1. Airplanes, Military—Evaluation.   2. Airplanes, Military-
-Piloting.   I. Aviation week & space technology.
TL685.3.M497   1995
623.7'46048—dc20   95-34729
                                                           CIP

Acquisitions editor: Shelley IC. Chevalier
Editorial team: Robert E. Ostrander, Executive Editor
                 Norval G. Kennedy, Book Editor
Production team: Katherine G. Brown, Director
                 Susan E. Hansford, Coding
                 Janice Ridenour, Computer Artist
                 Rose McFarland, Desktop Operator
                 Lorie L. White, Proofreading
                 Jodi L. Tyler, Indexer
Design team: Jaclyn J. Boone, Designer   GEN1
             Katherine Lukaszewicz, Associate Designer   0030898

Contents

Introduction

The pilot reports on military aircraft come from the pages of *Aviation Week & Space Technology* and span almost 20 years. The most recent pilot report on a military aircraft is mine on the U.S. Air Force's Northrop-Grumman B-2 that appeared in the April 17, 1995, issue. The earliest is a 1976 report on the U.S. Navy's T-34C trainer by the magazine's current editor-in-chief, Donald E. Fink. The Beech T-34C is still very much in use by the Navy as a primary trainer.

With only one exception, reports have been written by pilot/staff writers. The exception was the report written by British pilot/journalist John E. Fricker about the flight in a Chinese FT-6 fighter in service with the Pakistani air force. Some of the pilot reports were written by Herbert J. Coleman and David A. Brown, both now retired from the magazine. A number were by Robert R. Ropelewski, and Brendan M. Greeley, neither of whom are no longer with the magazine.

The pilots performing military evaluation flights primarily have operational military and not flight-test backgrounds. This is not by design, but dictated by the fact that our editors must be versatile and able to be journalists first and pilots second. William B. Scott, the Rocky Mountain bureau chief, is the lone exception in that he is a graduate of the U.S. Air Forces' flight test engineer course at Edwards AFB, Calif.

Ropelewski was a Marine Corps helicopter pilot with fixed-wing aircraft time, while Greeley was a Marine A-4 Skyhawk pilot. Coleman flew B-17s for the Army Air Corps in World War 2. I was a Navy Skyhawk pilot, but later flew Boeing 707s as a pilot/flight engineer for Pan American World Airways. Brown and Transport Editor Edward H. Phillips gained their flight time in private/commercial aircraft.

Because of the background of our editors and belief that while we have many test pilots reading the magazine, our general readers are more interested in overall performance and gaining a "feel" of an aircraft, we avoid inserting too much technical detail in the reports. The editors do not quote numbers just to impress the readers.

The magazine's policy always has been to try and be the first to fly any new aircraft and first in print with the pilot report. We have succeeded most of the time. Most of the reports have been generated by our editors by approaching either the aircraft manufacturer or service operating the aircraft. This process has been quick in some instances, but more often it has taken months and sometimes years to coordinate flights. There have been a few instances of manufacturers or operators telling us without prior warning that aircraft are available.

The operators attempt to give us an opportunity to plan for a flight, but at times a phone call has lead to quick packing of a flight suit and a hasty departure to wherever the flight is to take place.

In the cases of larger aircraft programs, such as the F/A-18, B-1 and C-17, the flights are preceded by simulator flights to acquaint the editors with systems and instrument panels.

The evaluation flights always are flown with experienced pilots, either a test pilot from the manufacturer or a service pilot highly qualified in the aircraft. Because of the time needed to qualify in an aircraft, insurance considerations and our general lack of current experience, the editors do not fly single-seat aircraft.

While some of the editors have used tape recorders to note data during a flight evaluation, I function best by using a Navy-issue kneeboard to record information. This information is melded with the results of thorough research of the aircraft's history, a preflight briefing and a postflight discussion to compile a pilot report. The editors have never promised a manufacturer anything but a fair and honest evaluation. Pilot reports are read by the magazine's own technical editors to ensure accuracy, but reports never are submitted to the manufacturer or operator for checking or tweaking prior to publication.

I have been asked numerous times which of my military evaluation flights was the most exciting. Most have been exciting, but for different reasons. When I became the first, at least known, rated Western military pilot to fly the Russian MiG-29, I had an experience I will never forget.

On a dark and overcast day, I flew the MiG-29 from the back seat—because of the weather—with Valery Menitisky. The back seat of the MiG-29 is not the best place for a pilot because of the limited visibility. But here I was, flying from Kubinka in Russia, reflecting on how the world and Cold War mentality were changing and that this flight would not have been possible even a few months earlier.

Eight months later, the excitement was of a different nature. In order to fly a different aircraft on each evaluation flight, one has to have attained a certain level of confidence. This flight was with Sukhoi's chief test pilot, Viktor Pougochev, in the Su-27 from Farnborough, England. This time, I felt my confidence may have done me in.

After performing an afterburner takeoff from the front seat, I was having trouble remembering that throttle latches must be lifted on Soviet aircraft to reduce power to the military setting. At the same time, I was attempting to talk to flight control using a radio control panel that did nothing for situational awareness, while converting meters to feet and kilometers per hour to knots in my head and flying in unknown airspace with a pilot in the rear seat who spoke little English. We were using the basic "Dave fly" and "Viktor fly." The flight went well despite the initial trepidation.

My latest flight, in the B-2, showed me that despite the somewhat unorthodox appearance of the stealth bomber, the aircraft performed like most, from the cockpit. This all was done using the magic of flight control system engineers and a digital flight control systems.

The view from the cockpit also reveals that aircraft are not aware of cost, program delays or potential political problems. This was true of the B-2, as well as the C-17 and many of the other aircraft flown by AW&ST editors.

David M. North
Managing Editor
Aviation Week & Space Technology

Chapter 1

Bomber aircraft

Control system key
to B-2 flight qualities

David M. North/Whiteman AFB, Mo.
April 17, 1995

On Mar. 22, Aviation Week & Space Technology *Managing Editor David M. North became the 61st person to fly the USAF/Northrop Grumman B-2. Of all who have flown the bomber since its maiden flight on July 17, 1989, North was the first who was not a Northrop or U.S. Air Force pilot, or a Defense Dept. official. U.S. Defense Secretary William Perry, Air Force Secretary Sheila E. Widnall and Under Secretary of Defense Paul Kaminski had recent flights in the B-2 at Whiteman AFB.*

The following articles on initial B-2 operations were written by North and William B. Scott, Rocky Mountain bureau chief, who accompanied North to Whiteman AFB. Scott has covered B-2 development, flight testing and production since the program's inception, and coauthored a book about the bomber's early days with then-USAF Col. Richard S. Couch: Inside the Stealth Bomber: The B-2 Story. *Senior Military Editor John D. Morrocco, who has closely followed B-2 developments in the Pentagon and Congress, also contributed.*

This report does not promote the B-2, nor does it advocate the building of more than 20 aircraft currently under contract to Northrop Grumman. That decision will be made later this year as the Air Force completes its bomber study and future requirements. Instead, the report is a comprehensive look at the bomber wing's accomplishments in operations, maintenance and training as it flies six of the 20 B-2s it ultimately will receive.

Both North, a former U.S. Navy pilot, and Scott, a former Air Force flight test engineer, were impressed with the positive attitudes and military sharpness observed among the wing's personnel. This professional attitude was prevalent—from the wing commander, Brig. Gen. Ronald Marcotte, to instructors, maintenance and line personnel. While there are still elements of the stealth bomber that could not be discussed, the wing personnel were as open and candid as possible.

The Northrop Grumman B-2 is more than satisfying the operational and maintenance demands placed on it by the U.S. Air Force during its first 16 months of flying from here.

The first B-2 was accepted by the Air Force's Air Combat Command on Dec. 17, 1993, and the 509th Bomb Wing launched its first training sortie five days later on Dec. 22. The bomber wing has accumulated more than 750 hr. since then and now has six B-2 Spirits in operation. The seventh aircraft is scheduled for delivery this month.

Whiteman will be the home of all 20 B-2s to be assigned to the 509th Bomb Wing. If an additional 20 B-2s are funded, they also most likely would be based here and assigned to the same wing, but within a second squadron, Brig. Gen. Ronald Marcotte, the wing commander, said. The base's support infrastructure has been sized to handle 40 aircraft, he said.

This *Aviation Week & Space Technology* pilot was scheduled to spend a day and a half with the 509th Bomb Wing—the first day to receive general briefings, followed by a preflight of the B-2. Later the same day, system briefings were to be held, and then time in the simulator to practice the next day's mission. The flight in the *Spirit of Washington* would take place early the next morning, followed by maintenance and pilot debriefings.

The B-2 preflight walk-around was conducted with the 509th's Col. William M. Fraser and Col. Greg Power, the vice wing comman-

Northrop Grumman B-2 bomber is powered by four General Electric F118-GE-100 turbofan engines.

B-2 is equipped with a center stick and throttles to the left of the pilots. The pilot sits in the left seat, while the mission commander occupies the right seat with easy access to data entry panel and cursor controller. B-2 pilots and mission commanders train as a two-person team, emphasizing crew resource management.

der and operations group commander, respectively. A preflight was done early because the following day the aircraft's engines would be running and all the checks prior to taxi would be completed.

The long, 172-ft. wing span makes a strong visual impact as a pilot walks toward the B-2 for the first time. Another notable feature is the clean, well-organized layout of the maintenance dock. Electrical power and cooling/heating were available from pits in the floor near the nosewheel well.

I also observed that there was no hydraulic fluid or fuel on the floor beneath the aircraft, or fuel seepage on the underside of the airframe. Power said that the 509th had not experienced any fuel or hydraulic leaks during its operations with the B-2. The lack of any fluid leaks was corroborated by Col. Henry Taylor, the wing's logistics group commander.

The overall smoothness of the airframe's surface also was apparent. There were no pitot tubes or other protuberances as found in conventional aircraft. There are four flush-mounted air data sensor groups that feed the four flight control computers.

One noticeable exception in the smooth airframe was what Fraser called "hair clips" in the forward edge of the wing that separates the leading-edge wing sections. Fraser said Northrop had designed a new process that would internally connect the sections so that the leading edge would be completely smooth.

Taylor said that supporting the requirements for low-observable materials is the biggest challenge in maintaining the B-2. He cited this as a relative challenge, because of the lack of problems in the aircraft's other systems, including the General Electric engines. The B-2 was designed to be maintained by opening as few access panels as possible. The Hughes radar, for example, can be accessed through the nosewheel well for almost all required maintenance. Taylor said the challenge is to establish the shelf life and requirements for low-observable material, such as tape and adhesives.

The current major inspection time for the B-2 at Whiteman is every 200 flight hours. The inspection and low-observable restoration occupies close to 44 workdays, although that time is expected to decrease as experience increases. The long-range goal for the major inspection is to increase time between overhauls to 400 hr. and then to 600 flight hours. The inspection interval to check for stealth degrada-

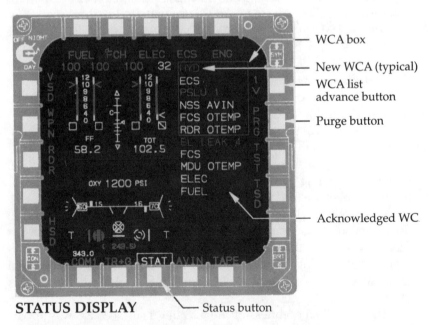

WCA box

New WCA (typical)

WCA list
advance button

Purge button

Acknowledged WC

STATUS DISPLAY   Status button

Maximum number of symbols shown. Not a typical display.

Status page shows fuel flow to the left and fan speed to the right in center displays. Elevon and rudder travel is indicated in the diagram to lower left. System functions are selected by depressing buttons on the display's outer ring.

tion is every 18 months. The first B-2 from the 509th is expected to fly over the radar cross-section ranges this fall to determine if there has been any degradation in stealth characteristics.

The pressure refueling connections are in the front wall of the left wheel well. A large digital counter on the maintenance dock's wall gives a readout of the fuel supplied to the B-2. There are two Halon fire bottles in each wheel well, which can be directed to either of the two F118-GE-100 engines or the Allied Signal auxiliary power unit. Boeing makes the main landing gear for the B-2, and it is similar to that installed on the 767.

With the landing gear down, the gust load alleviation system (GLAS) at the rear of the fuselage is deployed approximately 11-deg. down. During flight, the GLAS fairs into the fuselage and functions as an automatic pitch axis trimming surface. Fraser said the system appears to be effective, although I did not experience any strong turbulence during my flight.

The B-2's bomb bay is surprisingly large, considering the deceptive thinness of the overall aircraft. During an earlier visit in the day, I was able to observe six Mk.84 bombs located on a rotary bomb rack in the B-2 Weapon Load Trainer. The rotary launcher is capable of carrying eight of the 2,000-lb. bombs, and there was ample room behind the

B-2 refuels behind a Boeing KC-135 at approximately 255 kt. Pilots must compensate for airflow interaction between the two aircraft when in the refueling position.

bomb rack to carry additional bombs or other loads. Northrop claims a 32,000-lb. payload for the B-2, which would be the equivalent of 16 Mk.84s. The two rotary launchers are not counted as payload. The B-2 has been qualified to carry the standard range of conventional bombs. The carriage of the B61 and B83 nuclear weapons has been demonstrated, but the B-2 will not be qualified for delivery of these weapons until 1996.

The engine air inlet of most aircraft reveals the fan section of the engine and little else. But looking down the S-shaped ducting of the B-2 showed the fan section of the two GE engines as well as the top half of almost the entire 17,300-lb.-thrust engines.

A flight in the CAE-Link Flight Simulation B-2 simulator was with Maj. James Smithers, who would be my instructor pilot the following day. Smithers, and the rest of the 509th pilots, praised the simulator maker and Hughes Training System for how closely the simulator approximates the flight performance of the actual B-2. Having a full up-and-running simulator prior to first aircraft delivery also was considered a big advantage by the Whiteman pilots.

Smithers briefed me on the procedures to be used during the flight. I sat in the simulator's left pilot seat while Smithers took the right seat, normally occupied by the mission commander on operational flights. I performed the takeoff, and then climbed to 15,000 ft. Smithers then showed me some of the system's data pages on one of my four multipurpose display units (MDUs). We would not have as much time in the aircraft to look at these pages in flight.

The B-2's eight fuel tanks are displayed on the MDU with both a digital readout of quantity and a graphic presentation of the amount remaining in each tank. Fuel management is completely automatic with a manual backup. The aircraft's desired center of gravity can be chosen, so that fuel management will maintain the selected figure. On our actual flight, the center of gravity was near 33% at all times.

Cabin conditioning, equipment cooling and electrical and hydraulic systems can likewise be displayed on the MDUs. The display for the flight control system was more complex, but gave any parameter desired, and the location of the elevons, GLAS and rudders.

Rendezvous with a simulated tanker was next, and after finding the correct position behind the tanker, I was able to maintain my position fairy well. The actual plug in the next day would not be as easy. However, the final landing in the simulator was remarkably smooth and one of my best ever in a simulator.

The morning of my flight, the temperature was 58F with broken clouds at 10,000 and 25,000 ft. There was to be a slight left crosswind for takeoff and less than a 5-kt. headwind component. When Smithers and I arrived at the *Spirit of Washington*, the engines and systems had been running for 30 min. to assure that there would be no maintenance problems for our flight. A later modification to the B-2 will incorporate an Alert feature near the crew hatch that will prepare the aircraft for a more immediate flight. Our call sign was "Spirit Eight," and it was AV-11, or the eleventh B-2 built by Northrop.

We both entered the B-2 through the crew hatch at the left side of the cockpit with the aircraft in the maintenance dock. The ladder is self-contained in the aircraft. The cockpit is large enough to allow two or three people to stand comfortably behind the two pilot seats.

Provisions were made for a third crewmember, if it were determined the mission required an additional pilot or electronic warfare officer. A hatch was cut into the fuselage to accommodate the ejection of a third person, but there is no jump seat of any sort now installed. Marcotte said that operations in the 509th have indicated that a third person is not required, but it would have been beneficial during training flights to accommodate a second B-2 student pilot.

I took the left seat while Smithers strapped into the right seat. Both pilot positions are nearly identical, except that I had the gear handle on my side, and the weapon and navigation functions are more conveniently located on the mission commander's side. There is an emergency landing gear lowering switch on the instrument panel near the right seat, if the pilot were incapacitated.

The B-2's gross weight in the maintenance dock was 277,000 lb., which included 124,700 lb. of fuel. The starting fuel had been close to 130,000 lb., Smithers said. The interim flight gross weight limitation is 305,000 lb., but that is expected to be increased to 336,500 lb. after further flight testing. The higher than required fuel load was chosen so that another crew could take the aircraft on a low-altitude training flight without having to shut down the aircraft and refuel after we landed. Because of the maintenance reliability of the B-2, this is done often, Smithers said.

Fuel flow at idle for the F118 engines was 1,250 lb./engine, and the throttle position was at 17% of its maximum travel. Engine temperature was 430C, which is closely equivalent to exhaust gas temperature. Smithers completed a very short pre-taxi check. All of the checklists I observed appeared to be minimal.

I taxied the B-2 from the maintenance dock, using only a slight increase in thrust and the nosewheel steering through the rudder pedals. A low-gain setting on nosewheel authority was more than adequate to make a 90-deg. turn from the dock to the taxiway. Smithers said that even at higher gross weights and in tighter areas, the low gain was more than adequate. A higher gain setting for nosewheel steering is to be released later.

The B-2 taxied at 10-15 kt. to the active Runway 19 at Whiteman. The brakes were effective, as I had to slow the aircraft a few times to stay within the taxi speed. Taxi speed was represented by a digital readout on the left side of the vertical situation display.

I advanced the four throttles on the left console to maximum power and watched the fuel flow hover around 8,000 lb./hr./engine. The brakes were held for close to 10 sec. to monitor engine instruments. Once released, the B-2 accelerated quickly reaching the 139-kt. rotate speed 5,500 ft. down the runway. I raised the nose by the center stick to 10 deg. on the attitude indicator, and a climbout was started.

The Air Force opted for the stick in the B-2, much as it had for the Rockwell B-1B. Pitch response was positive, and if anything there was a tendency to want to raise the nose higher than the target attitude. Raising the gear did not change the pitch attitude. The 4,000-psi. hydraulic system raised the landing gear more slowly than I expected. But since I had flown the F-16 most recently, I was expecting a faster gear-transition rate.

The initial climbout was at 280 kt. to 9,000 ft. During the climb at maximum continuous power, the rate of climb was between 2,000

Deep bomb bay can carry 80 Mk. 82 500-lb. bombs or 36 cluster bombs. The planned range of the B-2 is over 6,000 naut. mi. with a 32,000-lb. payload.

and 3,000 fpm. The visibility from the left seat is excellent for a large aircraft. I was able to see the raised port-location light near the wingtip easily and, without stretching, the No. 1 engine inlet could be observed. The fine metal mesh built into the windows to deflect radar does reduce the visibility somewhat, but not to the point of being distracting. However, I did miss seeing an aircraft nose through the front window.

At 9,000 ft., the B-2 was leveled and the speed stabilized at 300 kt. The first maneuver any pilot would want to make in the B-2 would be a sharp banked turn. I raised the nose approximately 10 deg. and banked initially to 30 deg., and then to close to 60 deg. The B-2's roll rate is impressive for an aircraft of its size and approximates that of a heavy fighter. At the 30-deg. position, I was pulling 1.5g to get a steady turn rate, while at 60 deg. I was pulling the maximum-allowed 2g. There was no tendency for the nose to drop during the turn, and I was able to maintain altitude easily.

The stability designed into the B-2 allows the pilot to set the desired attitude and the aircraft will stay there. Also, the same stick forces maintain a 60-deg.-bank turn as well as shallower turns. One performance feature I noted throughout the flight was that deceleration was slow, even when flying at higher bank angles—an indication of low airframe drag.

The operational squadron is currently restricted to performing a maximum of 60 deg. bank turns. Smithers said the bomber is currently cleared to approximately 80% of its flight envelope and that the restrictions will be eased as it completes flight testing at Edwards AFB. However, the final bank-angle limit may not be much more than 60 deg. because there is not an operational requirement to go much beyond that in the heavy bomber.

Pitch response also was positive, but did not give the same impression of quickness as the roll rate. During the simulator flight at 10,000 ft. and 300 kt., I yanked the stick back and quickly received a stickshaker warning. There is a rate limiter tied to the angle of attack. The B-2 also has an angle of attack limiter that prevents the aircraft from exceeding its limit, similar to the system installed in the Lockheed Martin F-16.

I pushed the right rudder and watched the electronic yaw ball stay green until reaching 5 deg., when it turned red, indicating a flight limit. Opposite rudder input gave the same results. When I released the rudder, the overshoot in the opposite direction was less than 2 deg. and was quickly damped. This was true of sharp pitch and roll inputs and illustrated the heavy damping of the aircraft's flight controls.

The quadruple digital flight control system produced good control harmony, and the artificial "feel" in stick response was excellent. Stick forces are light but not sloppy in any way. The challenge to attain this control harmony in an aircraft without a tail has been more than well met by the flight control engineers.

One flight control system feature is control stick steering (CSS), which enables holding specified roll or pitch attitudes. If the bank angle is over 30 deg., the feature will return it to 30 deg. CSS also will reduce pitch to a 15-deg. attitude.

Deploying the speed brake elevons at 320 kt. while still at 9,000 ft. produced a slight rumble in the airframe, but with no pitch change. Deceleration to 200 kt. was rapid. The same stick forces and response to control-stick inputs were evident at the slower speeds as they had been at 320 kt. Engine response in increasing speed was quick with no lag, while acceleration at this mid-gross weight also was good.

At this point, I took a little more time to survey the instrument panel. Smithers was using his data entry panel to give me heading cues on my vertical situation display to stay within the Whiteman operating area. The horizontal display showed the operating area and would later be used to indicate the course to the waiting tanker. The left MDU contained the engine instrument display, while the right MDU contained the status display. Included on the status display are elevon and rudder displacement and fault warning announcements. The only fault warning we were to receive during the 2-hr. flight was of an intermittent IFF transponder.

The human engineering incorporated into the cockpit is evident. The stick and throttles are easy to reach and comfortable to operate. While the B-2 is designed to fly a mission to automatic bomb release and return on automatic pilot, those functions that you might want to do manually are easily accessible.

The B-2 does not have an automatic throttle system, so there is at least one major function required of the pilot—to maintain the desired speed. Most of the function display changes are done by pushing buttons located around the MDUs.

One display I was not allowed to call up was that of the Hughes AN/APQ-181 radar. However, the low probability of intercept J-band radar was given high marks for its fidelity and reliability by the 509th pilots. The terrain-following (TF) and avoidance feature of the radar is not yet available to the wing. Whiteman pilots are flying low-level training missions in the 1,000-2,000-ft. band until the TF is cleared.

One button on the eyebrow panel that I did not activate was the "penetration mode." This reduces elevon and rudder movement when maneuvering to minimize radar reflections during an attack. When the wing pilots were asked whether air traffic control is able to track them in the operating area, the answer was generally "yes." This is due to some radar-reflecting devices on the airframe and large control surface movements during training flights. In addition, not all planned stealth features have been incorporated into Block 10 B-2s.

Once the tanker from the 19th Air Refueling Wing at Robins AFB was located at 19,000 ft., I flew a rendezvous while the KC-135R was in a 30-deg. bank turn. I positioned the B-2 close behind the tanker, then Smithers flew a perfect approach for a hook-up. Because the tanker was loaded with JP-4 fuel and the B-2 uses higher energy JP-8, Smithers elected to not take on fuel.

During a second hook-up attempt, I had difficulty finding the correct formation box under the tanker. The B-2's inflight refueling receptacle is relatively far behind the cockpit; thus the B-2's wide planform is under the tanker, creating considerable airflow interaction between the two aircraft. I initially felt some reluctance about pulling the nose of the $600-million B-2 ($2.2 billion if you count program costs for 20 aircraft) up to the tanker. I finally did get a brief plug-in and decided that was enough.

Maneuvering behind the tanker at 255 kt. also revealed the B-2's tendency to go slightly nose-low with the addition of power and to pitch up with power reduction. Smithers said that in earlier flight control software packages, this tendency was more pronounced, and future versions will further reduce pitch changes with power. The fuel flow for the B-2 in the refueling pattern was 4,700 lb./hr./engine.

I pulled power to idle and banked 45 deg. for the return to Whiteman. At one point in the descent, Mach 0.80 was reached and control response was excellent. The noise level in the cockpit remained very low at the higher speed. The B-2 is touted as a high subsonic aircraft. Smithers said that when it is flown at 37,000 ft., for example, a Mach 0.78 cruise is used.

At a lower altitude, at 240 kt., with the power still at idle, the attitude indicator showed a 2-deg. glideslope. As we leveled at 8,000 ft. momentarily, I noted that the deceleration rate was close to 1 kt./sec., indicative of the aircraft's clean design.

While our profile did not call for high-speed, low-altitude flight, Smithers said the speed used on low-level training flights is 420 kt. and that the bomber is very stable in turbulence. Severe turbulence results in a longitudinal rocking motion, he said. The B-2 is stressed for high gust loads, a lot more than the 2-g limit would indicate, another B-2 pilot said.

The first landing at Whiteman was made using a coupled autopilot approach on the instrument landing system. Depression of the approach button on the eyebrow panel deploys the outboard rudders for a 90-deg. spread, using drag to allow better deceleration in the pattern. The landing gear was lowered below 224 kt. The autopilot was disconnected at 500 ft. with a speed near 145 kt. I compensated for the 9-kt. crosswind with a slight left wing low and right rudder.

Here is where it would have been helpful to see the aircraft's nose in order to judge yaw. The B-2 is not equipped with a headup display (HUD), which also would have aided me in the landing. While some of the wing pilots said they would like a HUD, it is not a high priority.

Power was reduced to idle over the runway chevrons and the landing was made on the centerline with a slight yaw at touchdown at 135 kt., producing a rumble in the landing gear. The rudders were retracted and power added for a go-around. Fuel flow in the pattern was near 4,400 lb./hr./engine. The next landing was made using a Tacan approach, with almost the same results, except that on this landing I made a slight flare and floated down the runway 1,500 ft. to a smoother landing.

On the next approach, Smithers demonstrated the ground effect on the B-2 by flying down the runway at approximately a 5-ft. altitude on the radar altimeter. The power addition at level-off was very small to maintain airspeed. My subsequent effort at the some maneuver resulted in a 10-15-ft. altitude.

My final approach resulted in a slight overshoot, which was corrected without being uncomfortable. The touchdown without flare at 133 kt. was one of the smoothest I have made in many years.

One comment on cockpit design is that the angle of attack indicator is to the right of center on the pilot's glareshield. On most left-hand approaches, I would prefer it on the left side. However, the landing airspeed is not on the backside of the power curve, so AOA is less critical.

Smithers made one more approach to a final landing and mentioned during rollout that the B-2's brakes and antiskid system were very effective.

I taxied the B-2 back to the ramp, where another crew was waiting to board the aircraft without shutting down the engines. Total flight time was 2 hr., with almost half that spent in the landing pattern. The fuel used from beginning taxi to return to the ramp was 35,300 lb.

During post-flight briefings with 393rd Bomb Sqdn. pilots and maintenance crews, the most often heard phrase referencing maintenance or material problems was that B-2 system specialists felt like the "Maytag repairman" with little to do.

The B-2 is yet another big step forward in the modernization of the U.S. Air Force's bomber force. While the Rockwell B-1 improved on the

Managing Editor David North enters the B-2 "Spirit of Washington" prior to his 2-hr. flight from Whiteman AFB with USAF Maj. James Smithers.

Boeing B-52, the B-2 represents a quantum leap beyond the B-1. The new technology embodied in the B-2 is reflected in the two-person crew, half that of the B-1 and later models of the B-52. The stealth characteristics of the B-2, combined with its range and payload capabilities, add yet another dimension to operational combat mission flexibility.

From a pilot's viewpoint, Northrop has given the Air Force an aircraft that is not only reliable, but easy and enjoyable to fly. While the B-2's unorthodox flyingwing design might suggest unusual handling qualities, the bomber flies much like other modern-day aircraft.

B-2 training pace exceeds expectations

William B. Scott/Whiteman AFB, Mo.

The Air Force's only B-2 bomber wing is about nine months ahead of its schedule to become fully operational, spurred by a growing fleet of reliable aircraft and hand-picked, highly motivated personnel.

Since receiving its first B-2 on Dec. 17, 1993, the 509th Bomb Wing has flown approximately 750 hr., achieving a 95% training sortie success rate. The unit now has six bombers, and will not achieve Initial Operational Capability (IOC) for another few years, but is rapidly expanding its reach beyond local training sorties. It already has participated in three power projection displays, and is refining its offsite deployment methods.

Sixteen operational pilots, including 12 instructors, are qualified to fly the bomber, and another eight are in training now. Air Force maintenance personnel are handling all flight line responsibilities—and have since the first aircraft arrived here. Contractor experts are still available and perform some specialized tasks, but Air Force 'bluesuiters' are more heavily entrenched than would typically be the case when a new weapon system is fielded.

The 509th Bomb Wing is flying more training missions earlier than had been originally programmed, primarily because its operational B-2s rarely require more than minor maintenance. Consequently, the unit participated in Red Flag combat exercises for the first time in February, even though procedures and operational concepts are still evolving. Eight missions were launched from here and flown over the Nellis AFB ranges of Nevada. All but one sortie recovered back at Whiteman after dropping two Mk.84 weapons on separate targets.

That B-2 landed at Nellis and changed pilots in what the unit calls an "engine-running crew change," or ERCC, a technique being developed as part of the 509th's operations concept. ERCCs are used routinely at Whiteman to squeeze two 4-hr. sorties into a single 8-hr. training period.

The Red Flag sorties consisted of four day and four night missions, and all flights were flown as scheduled within a four-week period. Inert Mk.84s were dropped during the day, and live weapons were delivered at night. All drops were made in the "bomb-auto" mode from level flight. Scores are classified, but all were "very good," according to a 509th senior officer.

"Weapons [delivery] displays and the radar are great. We're getting incredible weapon accuracies, and system workload is low," Maj. Richard Vanderburgh, a B-2 instructor pilot and training unit flight commander, said. "We'll win the war at night," because the bomber was designed to operate under cover of darkness, he noted.

The first 10 B-2s being delivered to Whiteman AFB are Block 10 versions that have a number of limitations but are adequate for training. They are both nuclear and conventional-capable, but only B-61/83s and Mk.84s, respectively, can be carried. Pilots and aircraft will not be nuclear-certified until Block 20 aircraft are available. For now, only fixed targets can be attacked until updated software becomes available.

Despite interim limitations, the six 509th B-2s have been remarkably problem-free—a testimony to an integrated logistics testing philosophy adopted early in the development program (*AW&ST* Oct. 12, 1992, p. 52). When the B-2's flight test plan was written in the 1980s, Air Force and Northrop leaders insisted that time be allocated for maintenance testing and technical order validation. Aircraft were scheduled for these "logistics tests," which were given equal priority with flight tests. That foresight is paying off now in high reliability and far fewer maintenance problems than is traditionally the case for new aircraft.

The initial suite of B-2 avionics also has demonstrated an unusual level of maturity, thanks in part to thousands of hours of testing in ground laboratories and on a C-135 airborne testbed. Radar and other B-2 systems have flown the C-135 for six years, developing coherent mapping, precision velocity updating and other features (*AW&ST* Oct. 12, 1992, p. 54). Air Force officials said most early radar problems experienced at Whiteman were corrected by a recent software update.

The first Block 20 bomber—AV-17—will be delivered to Whiteman in July, 1996. Although still short of complete mission-ready software, this version will be certified to deliver all conventional weapons planned for the bomber, and will include the GPS-aided targeting system and GPS-aided munitions (GATS/GAM).

Block 20 aircraft will have a limited defensive management system, some inflight replanning capability and only part of its terrain following/terrain avoidance functions. These features currently are still under development at Edwards AFB, Calif.

The full-up Block 30 capability will not be available until AV-20 arrives in August, 1997. All B-2s will start being retrofitted to Block 30 configuration about that time, adding a capability to carry joint direct attack munitions and other advanced conventional weapons. The B-2 also will be capable of locating and attacking "maximum defended" and moving targets. All "low observable" or stealth characteristics, a full defensive management system suite and terrain following/terrain avoidance features will be integrated into Block 30 aircraft.

Simulators enhance B-2 training efficiency
William B. Scott/Whiteman AFB, Mo.

Advanced, high-fidelity simulators and a reliable six-aircraft B-2 fleet are key contributors to efficient, accelerated flight crew training. Sixteen operational pilots at Whiteman AFB, Mo., are qualified in the bomber now, and three are cleared to fly conventional combat missions.

The first operational B-2 crews include a twelve-member Air Force instructor pilot cadre, 509th Bomb Wing senior staff officers and two OT&E pilots that transferred from the Combined Test Force at Edwards AFB, Calif. Another eight pilots are currently in training at the wing's Formal Training Unit here. Ultimately, about 42 mission-ready pilots will be needed to operate the unit's two squadrons of combat aircraft.

Instructors were checked out in the bomber faster than anticipated, even while simultaneously refining the training program. A well-integrated mix of academic, computer-based procedures, simulator and actual flight training—developed by Northrop, Hughes, CAE-Link and other subcontractors to meet Air Force criteria—is paying dividends in efficiency.

Another important factor has been the B-2's reliability. The wing has maintained a 95% "training sortie success rate" because "the airplane doesn't break," Col. Greg Power, 509th operations group commander, said. Before the first B-2 arrived at Whiteman, "we expected to fly about once a month. But now, we're flying four, six or eight sorties a week." That pace has enabled training more pilots in a shorter time than was anticipated, Power said.

Brig. Gen. Ronald C. Marcotte, the 509th wing commander, also observed that having functional flight crew and maintenance simulators before the first B-2 arrived "has been an unbelievable asset to our training program." From its inception, the B-2 program was structured to make sure training and maintenance resources would be available to the 509th when it started receiving aircraft (AW&ST Aug. 3, 1992, pp. 64, 67).

Class no. 4, the first class of full-time B-2 line pilots (not instructors), began academic training on Mar. 20. The class comprises four hand-picked students—two right-seat "mission commanders" and two left-seat "pilots," as B-2 crew positions are designated. Although both are experienced pilots, mission commanders (MC) are considered the senior crew member and are responsible for operating the radar, navigation and weapon systems. They also direct the mission and ensure either nuclear or conventional weapons are delivered on target. Pilots focus on flying the aircraft, but can drop ordnance if the MC is disabled.

Because most of the entire mission can be performed from either seat, both mission commander and pilot students receive almost identical training. The MC is given more "flights" in the Mission Trainer (MT), which is a specialized simulator focused on systems operation, not flying. Pilots typically will upgrade to MC status after about two years of B-2 line experience.

Pilots aspiring to fly the stealth bomber currently go through a rigorous, lengthy screening process that, so far, has eliminated about 90% of the original qualified applicants. Bomber pilots must have at least 1,000 hr. of flight time; fighter pilots need 600 hr. All must have 3-4 yr. of operational flying experience.

The initial instructor cadre consisted of only experienced bomber pilots, but the creation of Air Combat Command (ACC) opened B-2 seats to those with fighter backgrounds as well. Selectees now are coming from F/B-111, B-1B, B-52, F-15, F-16 and F-117 units.

After initial screening by ACC and 8th Air Force headquarters personnel, 15–20 prospective B-2 pilots are sent to Whiteman AFB for a week of scrutiny. Each pilot undergoes a skills evaluation flight in a T-38 Talon, a mock mission in the B-2 Weapons System Trainer (WST) simulator and an interview with Marcotte, the wing commander.

The names of primary and alternate selectees are forwarded by the 509th to 8th Air Force for approval, then on to the ACC commander. Gen. Mike Loh, the current ACC chief, interviews each pilot on the list and makes the final selection.

Of an original 85 qualified applicants for Class No. 4, four pilots were selected. Future classes may have 8-10 students, and up to four groups will be trained annually under current plans, according to Maj. Richard Vanderburgh, a B-2 instructor pilot. However, additional simulators may be necessary to support that many students per class.

Each basic class's Initial Qualification Training includes:

- About three months of academic courses that focus on B-2-specific systems and characteristics.
- Thirty-seven Cockpit Procedures Trainer lessons to familiarize pilots with system controls and develop proficiency in their individual duties. Most CPT sessions are 2 hr. long, but some can last 4-6 hr. Students can use the CPT on their own to reinforce certain training elements, calling up a computer-synthesized "surrogate pilot" mode to handle the other-seat duties.
- Specialized courses in the mission trainer. Designated left-seat pilots get two introductory MT sessions to better understand the mission commander's duties. MCs get five 2-hr. MT "flights." However, each MT session also requires an hour of preparation and an hour's critique.
- Fourteen missions in the Weapons System Trainer (WST). Six are 4-hr., systems-intensive sessions that also include five hours of planning and critiques. Each WST scenario takes 1.5 days to complete. Actual B-2 flights begin after about seven

WST missions are completed. Overall, each student gets 50-60 hr. of WST time.

- Three days of crew resource management training to foster teamwork. Developing individual expertise is the focus of cockpit procedures, mission training and academics phases, but CRM "pulls it all together in the WST" by emphasizing team performance, Vanderburgh said.
- Actual B-2 and T-38 flying. For proficiency, students and instructors fly the T-38 about 8-9 hr./mo. Instructors are getting 8-15 hr. of B-2 time each month, but students typically will fly only about 28 hr. during the 5.5-month course—six flights and a check ride. Each B-2 flight is preceded by a rehearsal mission in the WST, a technique that has proven highly effective in improving skills and retention.

"Most pilots view simulators with distaste, but the WST is [a very good representation] of the aircraft," Vanderburgh said. "After his first flight, [a pilot] looks back and says, 'Yeah, those WSTs were really worthwhile.' That first flight can be very overwhelming. I don't know of anybody who had a boring first flight [in the B-2].

"The WST, though, is definitely saving money and giving us a better pilot. We're seeing ideal transference of learning between the sim and the aircraft—tremendously improved ability to plan, to learn and grasp concepts," he said.

B-2'S future hinges on debate over cost

John D. Morrocco/Washington

The U.S. Air Force is pressing for a national debate on buying additional B-2 bombers which will begin in earnest later this month when the Pentagon releases an independent study of U.S. heavy bomber requirements.

Although Defense Secretary William Perry and other senior Pentagon leaders have stated they have no plans or money to buy additional B-2s, Congress left the door open last year when it called for a review of U.S. heavy bomber force requirements. The results of the independent review conducted by the Institute for Defense Analyses were to be briefed to Undersecretary of Defense Paul Kaminski last week, and he will in turn brief Perry. The study will then be presented to Congress before the end of the month. A second, related study on the bomber industrial base is due July 1 (AW&ST Feb. 13, p. 28).

With the Republican majority in the new Congress, support for the bomber is greater than last year, when pro-defense Democrats teamed up with Republicans to keep the Northrop Grumman B-2 production line alive through the end of Fiscal 1995. Republican lawmakers have continued to hammer at the mismatch between the Clinton Administration's military strategy and the resources it has provided to carry out that strategy.

Although the Pentagon's own Bottom-Up Review of defense requirements stated that the U.S. needs a force of 184 bombers to fight two major regional conflicts, the current budget plan only includes a force of 100: eight B-2s, 48 B-1Bs and 44 B-52s. The total would increase to 154 by 2000 as the remaining B-2s now on order are delivered and more conventionally capable B-1Bs are brought into the force.

Currently, the Pentagon is relying on a "swing" strategy for bombers if the U.S. is faced with two nearly simultaneous conflicts. At least half of the bomber force would be shifted from one theater to another. Gen. John M. Loh, head of Air Combat Command, labelled this unproven strategy a "risky" one. "I am not saying we can't do it." But he said it would put additional strains on Air Force personnel, logistics and airlift.

Loh told *Aviation Week & Space Technology* that long-range bombers are crucial in the opening days of a conflict, particularly in light of diminished U.S. overseas basing. Bombers would take on 85% of the targets in the first 10 days of a conflict, before tactical aircraft could arrive in large numbers in theater. Loh said B-2s would be targeted against the most difficult and time-critical targets, including airfields, air defense, command and telecommunications centers, and tactical ballistic missile launchers and storage facilities.

But while the Air Force preaches that bombers are central to U.S. global power projection strategy, service officials say they cannot afford to raid other accounts to pay for them. Loh told lawmakers that the Air Force does not want to buy additional B-2s at the expense of other high priority service programs, such as the F-22 fighter. And even if the service received a $1-billion windfall in new budget authority, Loh said there are "other higher priorities" than buying more B-2s, such as purchasing more F-16s and F-15s. Loh argued that if the bomber is the centerpiece of the U.S. global reach strategic policy, then it should be given the appropriate priority in terms of the notional budget.

The Air Force's insistence that the money come from sources other than its own budget will be a tough sell on Capitol Hill. Deficit reduction and tax breaks have overshadowed calls for increased defense spending in the Republican-dominated Congress (*AW&ST* Apr. 10, p. 20).

But the big difference from last year's B-2 debate is the apparent willingness of even the most deficit-sensitive Republicans to freeze the Fiscal 1996 defense budget at the current level of $270 billion. That would provide roughly an additional $12 billion over what the

Clinton Administration has planned for. According to Loh, roughly $750 million would be required in Fiscal 1996 to start buying 20 more B-2s at a rate of three per year. That figure would increase to around $2.6 billion in Fiscal 1997 and successive years.

Another factor working in favor of the B-2 is the shift in the defense debate away from concerns about readiness. Many lawmakers are worried that too much money is being cut from procurement and R&D accounts to fund readiness improvements. This backlash against bolstering near-term readiness at the expense of future readiness could work to the B-2's advantage. But the price tag may ultimately be too much for Congress to swallow.

During a recent hearing, House National Security procurement subcommittee chairman Rep. Duncan Hunter (D.-Calif.) ran a mini-tutorial on the different cost estimates for buying 20 more B-2s. Hunter had Air Force witnesses explain each of several different ways of calculating B-2 costs.

Northrop Grumman has provided the Air Force with an offer to build 20 B-2s at a total, nonrecurring flyaway cost of $12 billion. This would be for production at a rate of three per year. The company has also offered a second proposal for a production rate of one-and-a-half B-2s per year, which would result in lower annual costs but a higher total cost.

Loh said that the contractor's use of flyaway cost figures was "disingenuous." It does not include training and support equipment, management overhead, spare parts and associated military construction. These are included in the total program acquisition cost, which Loh said is "the real cost" to field and fly the weapon system. Air Force estimates also take into account contingent liabilities, sustaining engineers, engineering change orders and essential performance warranties not included in the Northrop Grumman proposal.

The Air Force has pegged the total program cost of buying 20 additional B-2s at a rate of three per year at $15.8 billion in constant Fiscal 1995 dollars. The total program cost at a slower production rate of one-and-a-half per year would be $19.6 billion.

But Rep. Ronald V. Dellums (D.-Calif.), a B-2 opponent, argues that the more relevant figure is what the taxpayers will ultimately have to pay to buy and operate the aircraft over its lifetime.

The Air Force estimates the cost to acquire and operate 20 B-2s for 20 years at $22.9 billion at a procurement rate of three per year, and $25.7 billion at a rate of 1.5 a year. Those figures are calculated in current year dollars, however. When adjusted for estimates of inflation, the figures are $21.5 billion and $36.5 billion, respectively.

B-1B combines brisk low-altitude handling, more capable avionics

David M. North/Dyess AFB, Texas
September 14, 1987

The Rockwell International B-1B displays fighterlike performance in the low-altitude mission and landing pattern, with excellent agility and stability. This is a quantum improvement from the Boeing B-52 it will replace as the Air Force Strategic Air Command's primary penetrating bomber.

A flight of more than 3 hr. by this *Aviation Week & Space Technology* pilot in the left seat of a B-1B from the Air Force's 337th Bombardment Sqdn. based here demonstrated the aircraft's low-altitude performance and landing characteristics, inflight refueling capabilities and the effectiveness of the offensive electronic systems. Without a specific threat against the B-1B during the low-level route over West Texas, the defensive electronic countermeasures system could not be fully evaluated.

Air Force Maj. Gary H. Flynt was the aircraft commander on the early morning flight. Flynt occupied the right seat of the B-1B, while I occupied the left seat alternating with Capt. Richard D. Davis, nor-

Visibility from cockpit is excellent in all flight regimes. Afterburner thrust on the four General Electric turbofan engines is used for takeoff in the B-1B, raising engine power from 17,000 lb. per engine to above 30,000 lb.

Cockpit configuration of the B-1B represents 1970s technology. The B-1B is equipped with sticks, rather than a yoke, and the throttles are positioned on the left side of the seats. Cathode ray tubes at each pilot station are used to display flight information and terrain-following data.

mally the copilot of the four-man crew. Maj. Steven J. Redmann was the offensive systems officer, while Capt. Robert M. Homola was the defensive systems officer.

Early fuel leak problem

Col. Albert D. (Don) Jensen, commander of the 96th Bombardment Wing, said the number of fuel leaks, an early operational problem in the B-1, had dropped from a high of 60 a month last October to fewer than 10 a month, with almost 60 aircraft delivered. A quick preflight inspection of the B-1B to be flown and several other aircraft on the flight line did not reveal any observable fuel leaks. As a former Navy pilot and airline flight engineer, I have observed many fuel leaks from commercial transports with more maturity than the B-1B.

The flight crew of Peyote 23, the call sign of the B-1B to be flown that day, had manned the aircraft prior to my arrival. The before-engine-start checklist had been completed prior to my arrival at the aircraft. Almost an hour was allotted between engine start and takeoff, time to allow the single Singer Kearfott inertial navigation system to align fully. Much less time could be taken for alignment, but in these early flights the maximum time is allowed. Air Force B-1s on alert status have external power connected at all times. The B-1B is equipped with two Garrett auxiliary power units, installed between the two pairs of engines, but during normal starts at Dyess, power is provided by underground cables.

Typical of problems that occur during the infancy of an operational aircraft was the inoperability of an electric motor that normally raises the crew ladder. The ladder had to be raised manually. Another motor was on order.

Engine start parameters

During engine start and initial taxi, I took the jump seat while Davis sat in the left seat. The four General Electric F101-GE-102 augmented turbofan engines were started at the same time. Flynt monitored the two engines on the right, and Davis watched the start parameters of the engines on the left. The engines reached approximately 700C on start and settled to an average of 520C at idle.

Fuel flow at idle was 1,000 lb./hr./engine. The only information that was classified during the flight was fuel flow at specific airspeeds. The B-1B has the capability of flying transcontinental unrefueled at its normal penetration speed of between Mach 0.85 and 0.90 at low altitude.

While I did not have the opportunity to taxi the B-1B to the takeoff position, other pilots who have operated the nosewheel steering said it handles very easily and precisely on the ground. Davis and I exchanged seats at the takeoff end of Runway 16 at Dyess.

The gross weight of the B-1B at the takeoff holding point was close to 348,000 lb., including 162,200 lb. of fuel. The weight at the ramp had been 352,000 lb. including 166,200 lb. of fuel. Close to 4,000 lb. of fuel had been consumed during the less than 1 hr. between engine start and takeoff roll.

Wing sweep of the B-1B was set at 15 deg., normal for takeoff and landing. The decision speed for the dry runway conditions at Dyess was calculated to be 159 kt. If the runway had been wet, the go-no-go speed would have been 139 kt. Rotation speed was estimated to be 174 kt., and 6,000 ft. was expected to be used for the takeoff roll. An approach speed of 184 kt. and a landing roll of 5,100 ft. would have been required if Peyote 23 had to return to Dyess immediately after takeoff.

Flynt was to make the initial takeoff in the B-1B, to allow me time to become accustomed to the instrument scan pattern. The B-1B is limited to an 8-deg. nose-up rotation attitude so as not to scrape the tail on liftoff. Many commercial aircraft are normally rotated to 15 deg. on takeoff.

All four afterburners were lit by Flynt, and the B-1B accelerated to its rotation speed in 4,500 ft., leaving 9,000 ft. of available runway.

Flynt said that at the lower gross weights used for training flights, the B-1B often was airborne in slightly over 2,000 ft.

I took command of the aircraft at an altitude of 4,000 ft. after the landing gear and flaps had been raised. The throttles were reduced to military power, without the use of afterburner. The wings were swept to 25 deg. for the climbout at 360 kt. and a 2,000-fpm climb rate. The wings are swept manually in the B-1B by the use of a large lever to the left of the pilot, or right of the copilot. Because of the intricate center-of-gravity considerations, an automatic wing sweep was not included.

The designers of the B-1B, however, did give Air Force SAC pilots other features associated with fighters, and not normally found in bombers. The aircraft is equipped with a stick rather than a transport or bomber yoke, and each pilot in the B-1B has the four throttles mounted to the left of his seat. These fighter-type features were especially appreciated during the low-level portion of the flight.

Less than 10 min. was used in the step climb to 20,000, mandated by air traffic control in the area. More than 10,000 lb. of fuel was used during the afterburner takeoff from Dyess and the climb to 20,000 ft. Fuel management in the B-1B is accomplished automatically and monitored on the front instrument panel. The current center of gravity and the outside c.g. limits also are displayed near the fuel flow and quantity gauges. Fuel is moved by highspeed pumps so that the c.g. will not exceed the limits at various wing sweeps.

During the time spent orbiting at 20,000 ft. until our low-level clearance, I was better able to observe the cockpit instrumentation and layout. The B-1B airframe and four-man cockpit contains a mixture of older and newer systems, a heritage of its checkered past.

The newest addition to the Strategic Air Command's aircraft inventory has had a tumultuous history. The B-1 bomber program was launched in 1970, but then canceled by President Jimmy Carter in June, 1977, after a long congressional bout over the role of the manned bomber and escalating costs. The Reagan Administration revived the B-1 program in October, 1981, and the first production B-1B flew in October, 1984. The bomber that has emerged from this stop-and-go past has a cockpit best identified with aircraft developed in the 1970s. Many current commercial transports and business aircraft have flight instrumentation surpassing that of the B-1B. Both the Air Force and Rockwell made the decision that new aircraft systems would not be incorporated into the B-1B unless they contributed directly to its primary role as a penetrating bomber. A head-up display for the B-1B was not seriously considered by the Air Force. While the HUD would have been an asset for low-level training, the system would not be used with the radiation glareshield in place during an actual mission.

Other systems, such as a second Singer Kearfott inertial navigation system, was omitted because of cost constraints. When the Air Force contracted for the 100 B-1Bs in 1982 for $20.5 billion in 1981 dollars, a number of systems were put on hold, with hopes of adding them later. Provisions have been incorporated into the B-1B for a second INS, including the wiring and warning display on the master caution light panel. The reliability and accuracy of the single Singer INS has proved to be good, with less than a 200-ft. drift in 3 hr. of flying. The mission dependency on the INS, however, dictates that a second system be installed.

The Air Force also is looking at the possibility of installing a forward-looking infrared (Flir) system in the B-1B as the need is better defined and money can be made available.

I found the tape displays used in the B-1B difficult to include in a good instrument scan at first. Angle of attack, speed in Mach and speed in knots are located to the left of the center cathode ray tube. The Sperry CRT provides heading information at the top and pitch and roll in the center of the display. Rate of climb, altitude and radar altimeter information is located to the right of the center CRT. All of the engine instrumentation is contained on tape displays, making engine parameters easy to read. Many of the aircraft's system displays are located on the overhead panel.

While observing the instrument displays and talking with Flynt, I often strayed from my assigned altitude. I was reminded of this by Redmann, the offensive system officer. When correcting back to 20,000 ft. from a lower altitude, the stall warning system often would sound. The stall warning system is programmed to sound at near 20% above stall speed. The installation of a stall inhibitor system, now under development, will lower the stall warning to 5% above stall speed.

Pitch-up tendency

As I slowed the aircraft at 20,000 ft., the stall warning sounded at 250 kt. at a 5-deg. angle of attack. The B-1B when flown into the stall and beyond has a tendency to pitch up and then fall off on a wing.

A feature in the B-1B that I discovered needed more time for adjustment was the sensitive pitch control in all flight regimes. I had a tendency to overcontrol the aircraft in pitch attitude at first, but this decreased as time passed at the controls. Roll control at altitude with 25 deg. of wing sweep was much less sensitive, a feature that Flynt said was built into the hydromechanical flight control system.

On command from Redmann, I nosed the B-1B over to arrive at the initial low-altitude point on time. The wings were swept back to

65-deg. at 11,000 ft. in the descent. During the wing sweep, there was little change in pitch.

The fuel at the initial low-level point was 133,000 lb. and the gross weight close to 318,000 lb. Approximately 2 hr. had elapsed since engine start and 1.3 hr. since takeoff. The general terrain elevation in the area was 4,500 ft., but there were many hills that went above 7,500 ft. on the initial legs of the low-level route. Altitude to be flown during the 45-min. route was 500 ft. over the hilly portion, dropping to 400 ft. when the terrain became flat.

The B-1B flown on the evaluation was equipped with the terrain-following system that uses the Westinghouse AN/APQ-164 multimode offensive radar for input. The aircraft also is equipped with low-observable phased array antenna for the terrain-following system. The terrain-following system could not be used on this flight, because Flynt and his crew had not received the training required. Flynt did, however, turn the system on during one low-level portion of the flight to observe the display.

The terrain-following system in the B-1B has greater than a 10-mi. range and is displayed on the lower segment of the pilot's CRTs. The terrain is shown in a horizontal format, looking at a plan view of the altitudes. Flynt said it is a better system than flown in the B-52 and the General Dynamics F-111. The radar has the ability to pick up a radio tower, and the terrain-following system will fly the aircraft 175 ft. over the tower, he said.

Earlier restrictions against using the automatic terrain-following system have been lifted, except when encountering problems in the system. The earlier restriction included use of the defensive electronic countermeasures system. During initial flight and operational testing of the terrain-following system, there were unwarranted pop-up alarms, causing the aircraft to pitch up when not needed. The Air Force has modified the computer software program for the terrain following, and there have been fewer unwarranted pitch-up commands. The lower training altitude would be 200 ft. with an operational system.

The B-1B was accelerated to Mach 0.85 at the 500-ft. level. Heading commands for the flight were provided by Redmann and displayed by a heading cue located above the directional display on the pilot's CRT. I found the roll control of the B-1B excellent with the wings swept at the low altitude. Pitch control at the same altitude appeared to be slightly dampened from that at the higher altitudes. To use terrain masking to hide the aircraft from expected defensive positions, wing banks of up to 60 deg. were used for larger heading changes.

The visibility from the cockpit was excellent at low altitude, both forward of the nose and to the side. The overall impression of the

B-1B's performance at low altitude was that of a fighter, and it was difficult to believe that there was almost 140 ft. of aircraft behind the pilot's seat. A two-section tailplane that moves in different directions during a turn and four-segment spoilers/airbrakes provide crisp roll control. There are no ailerons on the bomber.

Smooth ride quality

Another feature of the B-1B that was noticeable at over 540 kt. and at 400 ft. was the smoothness of the ride. During the entire low-level portion of the flight at near noon in West Texas, there was little turbulence. During the portion that Flynt flew the aircraft, I found that it was easy to make notes on a kneeboard. The B-1B is equipped with a stability control augmentation system to help eliminate the effects of turbulence.

The coordination among Flynt's crew was excellent. Callouts of terrain and other obstacles by Redmann, who monitored the Westinghouse radar display, were timely. Redmann also alerted us for impending release points for various simulated weapons that were to be released during the low level run.

Releasing the simulated short-range attack missiles on the B-1B required a stable attitude and a slight climb. Conventional weapons and nuclear weapons also were dropped in simulated runs, and all weapon releases were monitored and scored with excellent results.

The defensive systems officer, Homola, relayed messages of simulated missile launches against the B-1B during the low-level portion

Strategic Air Command B-1Bs on the flight line at Dyess AFB are part of the 96th Bombardment Wing and are used for training and operational missions. The initial training for all B-1B crewmembers will remain at Dyess.

Rockwell International B-1B was designed to operate at speeds of up to Mach 0.9 at altitudes near 200 ft. in the penetrating role. The B-1B was found during an Aviation Week & Space Technology *flight to be highly maneuverable at these speeds and very stable in the low level environment.*

of the flight. These simulated threats included Soviet-built SA-3 and SA-10 surface-to-air missiles. The reports contained the location of the firings. At the same time, Homola would use the appropriate countermeasure, such as flares or chaff against the missiles. Changes of direction also were flown to best counter the threat.

Soviet radar potential

Development of the Eaton/AIL AN/ALQ-161 defensive avionics system has been the most challenging problem for the B-1B program and has delayed the bomber's full capability in this area for several more years. The full realization of the potential of the Soviet threat, such as full look-down/shoot-down radars, has extended the development program for the ALQ-161. Additional requirements levied on the system, plus the inability of the defensive management system to automatically detect and counter Soviet type threats at a high level of reliability and repeatability, have slowed development of the full potential of the system.

The Strategic Air Command depends on the aircraft's ability to fly at low altitudes and at high speeds as one of its defensive factors. Another defensive asset for the B-1B is its low radar cross section. The B-1B has 1/100 the cross section of the B-52 and one-tenth that of the B-1A. The radar cross section also is one-seventh that of the FB-111.

"First they have to find us, and then the fighters have to come down low to get us and we will run them out of gas," Lt. Col. John Wilcox, commander of the 337th Bombardment Sqdn., said. "At least the early returns from Air Force Red Flag exercises indicate that is happening. In order for the monitoring radar sites to get our position, we either have to climb or put our transponders on."

At one point near the end of the low-level route, the B-1B was slowed to 430 kt. With the use of afterburners the speed was increased to 555 kt. in less than 30 sec. in level flight. The stability and smoothness of the B-1B at 400 ft. was still impressive to this pilot, who had flown the McDonnell Douglas A-4 on many similar missions as a Navy pilot.

The low-level portion of the flight was completed after 45 min. The gross weight of the aircraft was now close to 281,000 lb. and the fuel on board was 95,200 lb. All four afterburners were lit at 400 ft. and at Mach 0.80. The aircraft accelerated to Mach 0.91 while I was still raising the nose of the aircraft to a 20-deg. climb attitude. The aircraft quickly exceeded our 17,000-ft. assigned altitude, and I used a modified 60-deg.-bank wingover to descend back to 17,000 ft. The Air Force has experienced little difficulty with the General Electric F101 engines during initial operational flying. A few engines have had foreign object damage, but the overall reliability has been excellent, Flynt said.

Refueling tanker position

Davis and I exchanged seats for the inflight refueling. The Boeing KC-135 from the 917th Air Refueling Sqdn. was spotted at a distance of more than 20 mi. at 20,000 ft. Flynt made the first approach to the tanker and had no difficulty achieving a plug-in the first time. The B-1B is refueled through a receptacle in the nose of the aircraft, making it easy for pilots to monitor the plug-in. The boom was then disengaged from the nose of the B-1B, and Davis flew an effortless approach and engagement to the tanker.

Flynt retarded the No. 1 engine to the idle position and flew into a reengagement with the tanker's boom. The remaining three engines were flown in the mid-range on the throttle quadrant. Placing the other left engine throttle to idle required Flynt to use intermittent afterburner to approach the tanker and stay in position with the two right engines. There was little yaw encountered during the refueling, and Flynt was easily able to counter yaw with rudder input.

Approximately 40,000 lb. of fuel was transferred from the tanker to the B-1B during the almost 30 min. behind the KC135. The fuel transfer could have been much faster had the B-1B remained engaged the whole time.

Landing demonstration

A descent was made from the refueling altitude to Dyess AFB, while I took the left seat again. The first landing at Dyess was made by Flynt to demonstrate the height of the eye above the runway at touchdown, always a difficult distance to gauge for the first time in a new aircraft.

Instrument landing system approaches were being made to Runway 16. The wind was from 180 deg. at 20 kt. The approach speed was 174 kt., with flaps and landing gear extended and the wings in the 15-deg. sweep position. Gross weight of the aircraft was close to 300,000 lb. Flynt's landing was smooth and a go-around was initiated.

I took the controls on the wide downwind for another ILS approach. There was some trepidation on my part in making the landings. The previous day, I had crashed twice in the nonmovable simulator while attempting ILS approaches starting from a 2,000-ft. altitude. I had difficulty correlating control input with responses mandated by unfamiliar glideslope symbology displayed in the center CRT.

Attempting to follow the aircraft symbol in the ILS rectangle still caused me to overcontrol the B-1B in pitch, and less in roll. Speed control during the approach was not difficult. Once I transferred to a visual reference with the runway, the tendency to overcontrol in pitch decreased.

A rate of descent of 700 fpm was held on the approach, with ground effect lowering the descent to 300 fpm at touchdown. The first landing was firm, but there were no bounces. I advanced power for the takeoff, and the aircraft showed itself to be very agile during a tighter pattern for the second approach and landing.

Takeoff with afterburner

Pitch control was less of a problem on the second approach, as I was more able to assimilate input to responses required for the ILS approach. The touchdown was smoother and another takeoff was made using the afterburners. As in low-level flight, the overall impression of the B-1B in the traffic pattern is that of a high-performance fighter.

Flynt demonstrated the B-1B's agility by maneuvering into a tight overhead pattern with a close downwind for a fighter-type approach.

Davis then took the left seat for an ILS approach and landing. His approaches were smoother than mine. This time, I observed the radar display from Redmann's position. The resolution of the radar was excellent, and Redmann pointed out a Boeing B-17 near the entrance to the main gate and the lamp posts leading into the main gate. The metal fence surrounding the area also was visible on the CRT display.

The final landing was accomplished, and Flynt stopped the aircraft in less than 5,000 ft. without heavy use of the brakes. He said the antiskid system is excellent and has the ability to stop the aircraft in short distances without overheating the brakes.

The B-1B was taxied to the ramp with a gross weight of 278,300 lb. and a fuel load of 92,500 lb. Elapsed time from engine start was

5.3 hr., and total airborne time was 4.3 hr. The total amount of fuel consumed during the 5.3 hr. was close to 114,000 lb.

Another crew exchanged seats with Flynt's crew with the engines still running. The new crew was taking the B-1B for inflight refueling and then through a low-level route in Colorado for automatic terrain-following training.

The B-1B may have had a difficult past, with political and development problems, but it is a large step forward in performance and ability from the B-52 it is replacing.

USAF/ROCKWELL INTERNATIONAL B-1B SPECIFICATIONS

Powerplant

Four afterburning General Electric F101-GE-102 turbofan engines with 30,780 lb. of thrust each.

Weights

Maximum takeoff weight	477,000 lb. (216,630 kg.)
Maximum payload	125,000 lb. (56,700 kg.)
Maximum fuel load	195,000 lb. (88,450 kg.)

(Additional fuel can be carried in the bomb bays)

Dimensions

Length	145.8 ft. (44.4 meters)
Height	34 ft. (10.4 meters)
Wing span, 15 deg. sweep	137 ft. (41.8 meters)
Wing span, 67.5 deg. sweep	78 ft. (23.8 meters)

Performance

Maximum speed	Mach 1.2 to 1.3
Penetration speed	Mach 0.85 to 0.9
Range at low level	Intercontinental can be refueled by KC-135 and KC-10 tankers
Takeoff roll at 348,000 lb.	4,500 ft.

Chapter 2

Fighter/attack

AMX fills air-to-ground role with room for mission growth

David M. North/Turin, Italy
July 15, 1991

The AMX has all the attributes required of a ground attack/reconnaissance aircraft along with the ability to perform in an air-to-air role and the potential for mission growth. The Italian/Brazilian AMX, had it been called to duty in the Persian Gulf war, could have performed credibly with its air-to-ground capability and internal electronic countermeasures systems.

During the early development of the AMX, Embraer's then-managing director, Ozires Silva, said the aircraft was being designed as a new technology replacement for the McDonnell Douglas A-4. The intent was to build a relatively simple, highly reliable attack aircraft at a reasonable price. The AMX also had to carry a large payload and be agile at high subsonic speeds at low altitude.

The AMX program was started in 1980 by three companies, Aeritalia and Aermacchi of Italy and Embraer of Brazil. Aeritalia, now Alenia, is the project leader for the AMX and is responsible for 46.5% of the workload. Aermacchi has 23.8% of the aircraft and Embraer 29.5%. The AMX is in service in the Italian and Brazilian air forces. The Italian air force has orders for 238 AMXs and the Brazilian air force for 79 single and two-seat aircraft.

Last month this *Aviation Week & Space Technology* pilot had the opportunity to evaluate the AMX from the Alenia test flight facility at the Caselle airport, near Turin. The flight was with Napoleone Bragagnolo, chief test pilot for Alenia.

Following a short briefing, I strapped into the front seat of the AMX. The attack aircraft is equipped with a Martin Baker Mk. 10L ejection seat. Ejections from the AMX are through the canopy, with the rear seat going first in the two-seat configuration.

Bragagnolo gave me a thorough front seat checkout in the aircraft prior to taking the back seat. The cockpit instrumentation in the back seat of the AMX is almost identical to the front seat layout, including a repeater of the head-up display in the front seat. This option could be dropped if operational flights do not prove the value of the HUD repeater.

The auxiliary power unit and engine start switches were in the front seat console only. The APU was started first, taking over internal power from the external power cart. The single Rolls-Royce Spey

AMX wing sections and air intakes are built by Embraer in Brazil and airfreighted to Turin. Aeronautica Macchi produces the nose and aft fuselage sections. Alenia has 46.5% of the workshare.

RB168-807 turbofan engine started easily by depressing the engine start switch and bringing the throttle to ground idle once the start light had illuminated.

The weight of the two-seat AMX was approximately 22,000 lb. on the ramp, including 4,850 lb. of internal fuel. The single-seat AMX has an internal fuel load of 1,100 lb. more, or 5,950 lb. The AMX is capable of carrying four external fuel tanks with in excess of 2,000-lb. capacity each on its wing stations for ferry flights. Two AIM-9L Sidewinder air-to-air missiles are carried on the single- and two-seat AMXs on wing tip launchers. During development of the AMX, placing air-to-air missiles on under-the-wing launchers was evaluated, but the wing tip location was found to be more effective.

I taxied the No. 7 AMX from the ramp to the active runway at Caselle. The high-gain nose wheel steering was activated by a switch on the stick and allowed a 45-deg. arc from the centerline. Nose wheel steering through the rudder pedals was effective. There was no tendency of the nose wheel to shimmy or of the wheel brakes to grab during the taxi. Bragagnolo said it took the developers seven tries before they reached the present nose wheel configuration, which he praised as being very effective.

In the two-seat prototype we flew, there is a detent in the idle range of the throttle. On the ground, a 53% high-pressure compressor power setting is used for idle, while in the air 60% represents idle power. On production aircraft, the idle setting is automatic.

The throttle was initially advanced slowly toward full power, until Bragagnolo said to jam it forward to judge engine response. The RB168-807 engine is not equipped with an afterburner. The brakes were held until power passed through 90% prior to release. I rotated the nose of the aircraft to about 13 deg. at 115 kt., and the AMX lifted off at 120 kt. The landing gear and flaps were raised while indicating 150 kt. Takeoff roll was slightly less than 2,000 ft. Maximum takeoff weight for the single-seat AMX is 28,600 lb.

The climb to altitude was made at 300 kt., transitioning to Mach 0.7 at the higher altitudes. During the climb out under air traffic control, I had the opportunity to spend more time looking at the instrumentation in the AMX. The front cockpit includes one multifunction display on the left front panel. The display primarily was used to show navigation points and entry into the flight test area during the evaluation flight. The display also can be used to show the armament load and the avionics system. Check lists and other aircraft systems schematics were not incorporated into the multifunction display.

A semi hands-on-throttle-and-stick (HOTAS) concept has been used in the AMX. The communication, navigation and IFF controls are still on the side consoles, but if a wide-view head-up display is added in later aircraft, an up-front control panel for those functions also will be installed.

After flying a wide-angle head-up display in the Israel Aircraft Industries Lavi demonstrator earlier this year, I found that I had to concentrate on using the smaller HUD in the AMX for flight information (AW&ST Mar. 25, p. 46). The wide-angle display is more natural to use in visual flight conditions. The AMX partners are looking at the possibility of installing a wide-angle HUD in the 1994-95 timeframe.

The display symbology in the attack aircraft's HUD, however, was excellent, and I found it easy to use for flight reference and later for simulated weapon releases. The display was bright, and there was no difficulty in reading the figures on the HUD at all times.

Climbing through 11,000 ft. at 300 kt., the rate of climb for the AMX was close to 6,000 fpm. I leveled the AMX at 24,000 ft. Bragagnolo said that because the AMX was designed for low altitude, it did not make much sense to fly at higher altitudes except on a ferry flight. The AMX has been flown to 44,000 ft., but its maximum operational altitude is 42,500 ft. The optimum high altitude is 26,000 ft., Bragagnolo said. As an example, at 26,000 ft. and a speed of Mach 0.77, the fuel flow of the single Rolls-Royce engine is 2,380 lb./hr. At 37,000 ft., the fuel flow is 2,025 lb./hr. flying at Mach 0.75. Not included in the calculations would be the fuel required to climb from 26,000 to 37,000 ft.

The combat range of the AMX on a low-level mission with only internal fuel and carrying six Mk.82 bombs, two AIM-9s and guns and ammunition is greater than 250 naut. mi. With fuel tanks that are dropped when empty and carrying 4,000 lb. of ordnance on a high-low-high mission, the combat radius is greater than 650 naut. mi.

At 24,000 ft., I retarded the throttle to idle. The altitude was maintained and aerodynamic buffet came at 130 kt. with the angle of attack slightly above 20-deg. The AMX did not roll off on a wing; instead the rate of descent increased at a slightly high nose angle until the nose was dropped and power added. The maneuvering flaps were elected next. Using 75% power and holding altitude, the AMX stalled at 113 kt. and at a 25-deg. angle of attack. Again, the stall was very benign, and I kept the same attitude until 800 ft. of altitude was lost before adding power.

Bragagnolo said that during the spin testing of the aircraft, the AMX was found to be spin resistant in the clean configuration, and departure resistant with maneuvering flaps. In most cases, if the controls are released after departure from controlled flight, the aircraft will recover from the spin after one turn, he said. The test pilots were unable to get the AMX into a flat spin, even after a sequence of five spins. Yaw rate during the spins was 110 deg./sec., and the roll rate was 200 deg./sec., Bragagnolo said.

The AMX partners are developing a flight control computer that will enable automatic deployment of maneuvering flaps. By the end of the year, the flight control computer is to be tested and installed in new aircraft. At the same time, the AMX will be equipped with a basic autopilot with only altitude hold and heading and track acquisition and hold, Gian Mario Avagnina, Alenia's program manager for the AMX, said after the flight.

The third batch of AMXs is to be equipped with the autopilot and will feature an auto approach mode with microwave landing system to Category 2 limitations.

While at 21,000 ft. and still with maneuvering flaps, slow flight was performed at 158 kt., while making 45-deg. bank turns. The AMX was stable and the nose did not wander. At 15,000 ft., I rolled into 45-deg. bank turns and performed some 180-deg. turns while pulling 4g. Again, the AMX was stable, and there was little buffet during the turns.

I then performed a split-S maneuver to reach a lower altitude, where the AMX was designed to be at its best. At the low altitude, the speed for best range is 300-360 kt., and the dash speed used was 450-500 kt. The speed limitation for the AMX is 550 kt. below 10,000 ft. I was briefed to fly at 500 ft. above the terrain, except over towns where the minimum was to be 1,000 ft.

Having flown the A-4 Skyhawk for in excess of 2,000 hr.—with much of that time devoted to flying low-level navigation routes and low-altitude weapon delivery—I was comfortable flying the AMX at these speeds and altitudes.

My immediate impression of the AMX in this element was of the excellent visibility from the front cockpit in all directions, especially over the nose, when looking for simulated targets. My next impression during the warm day over the hilly terrain south of Turin was that the gust response also was excellent, providing a very stable platform.

At one point, I was flying at close to 450 kt. and fuel flow was less than 3,700 lb./hr. Engine response was excellent while trying to maintain 450 kt. and making sharp turns to stay above a valley. The visibility from the cockpit also was helpful while making 5g turns during the low-level run.

The turns were done to simulate the detection of an airborne attacker from the rear. The AMX has an internally mounted electronic warfare package, including active electronic countermeasures. There also is an infrared missile launching and approach warning receiver and a radar warning system, both for the front and rear hemispheres. The chaff and flare dispenser system is automatic and contained internally.

The air-to-ground and attack mode was selected during the low-level, and a building in the distance was chosen as the target. The building was selected by using the designation button on the throttle to move the target cross-hair on the HUD. The range-only radar gave a distance, and the radar altimeter was used for altitude. When the solution was right, a simulated bomb was released automatically.

The primary mode used in the attack was continuous computed impact point. After flying over the target, a steering bar was displayed on the HUD to indicate the return to the same target. I flew out about 10 naut. mi. before making a run in from a different direction. This time a toss release was selected, and during the run in, a command to climb was shown on the HUD with a bomb release at 32-deg.

All during this low-level run, the altitude above the ground never was above 700 ft., and the speeds were from 400 to 500 kt. The pilot workload for weapon delivery was not heavy. Because I was not familiar with the system, I had to search for the appropriate HUD symbology and weapons switches.

Bragagnolo said that the performance of the AMX at low altitude is not degraded appreciably with the addition of a true weapon load, such as six Mk.82s. The wing loading of the clean AMX is 400 kg./sq. meter, and with six Mk.82s, the wing loading is 520 kg./sq. meter. The ability to sustain a high maneuvering rate is downgraded with the weapon loads.

The AMX partners have looked at installing a higher-thrust engine in the AMX to achieve a higher sustaining turn and climb rate, but the

Cockpit of the AMX includes one multifunction display used primarily for navigation and weapon information. A moving map display capability is to be added. The rear cockpit in the two-seat AMX is nearly identical to that of the front seat, including a head-up display repeater.

Italian and Brazilian air forces do not require it for their missions. For an export customer, AMX International would consider installing a General Electric F404 engine, with or without afterburner. However, the Rolls-Royce engine has a 13% lower specific fuel consumption than the higher-thrust F404 with afterburner and an infrared signature that is 70% less, Avagnina said.

In the air-to-ground mode for firing rockets and guns, lateral wind is compensated for in the inertial navigation system, so the pipper is placed on the target without any offset. The Litton inertial navigation system has been very accurate to date, Bragagnolo said. A second INS or a Global Positioning System is being evaluated for later aircraft as a backup.

After the simulated ground attack, the air-to-air mode was selected, and a gun pipper was shown in the HUD. A false reading on the radar gave us a target and showed the radar range ring around the pipper. The hot line of bullets also was shown. Bragagnolo said the gun system is very accurate, and the bullets will go where the pipper is placed. The Italian AMXs have one internally mounted 20-mm. Vulcan gun. The Brazilian version has two 30-mm. DEFA guns mounted internally.

At 5,000 ft., an acceleration check was made. It required close to 22 sec. to accelerate from 300 to 450 kt. in the clean configuration. Also at this altitude several aileron rolls were performed. They were done slightly nose high at 320 kt. with a roll rate of 170 deg./sec. The stick was not fully deflected to achieve that rate. A rate of 230 deg./sec. has been recorded in flight test, but a 240 deg./sec. rate is the design maximum.

I then pulled 5.5g to enter a loop at close to 5,000 ft. at a speed of 380 kt. The speed on top was 170 kt., and on the recovery the speed was 310 kt. The maximum g-loading for the AMX is 7.33. Other acrobatic maneuvers were performed with no surprises.

During the flight back to the Caselle airport, I asked Bragagnolo whether a side-stick controller had been considered for the AMX. He said that it had, but because the primary mission of the AMX was attack it was felt that all aircraft systems should be reliable and be able to suffer battle damage and still perform the mission. The flight controls are operated by a dual hydraulic system, and there is a manual backup system. The manual backup system would not be available with the side-stick controller, Bragagnolo said.

The AMX partners also have shielded the critical mission equipment with nonessential equipment. The avionics, electrical and fuel systems also have redundant systems to ensure survivability. There has been one recent crash of an Italian air force AMX. Bragagnolo said that it appeared that the primary fuel monitoring system failed in

The AMX is capable of carrying a wide range of weapons. The outboard wing pylons have a load capacity of 1,000 lb. each while the two inboard pylons can carry 2,000 lb. each. Two Sidewinder AIM-9Ls are carried on the wing tip stations. A twin bomb rack can be loaded on the fuselage station. An internally mounted camera pallet can be installed.

the AMX, leading to an engine flameout. The pilot ejected safely, but did not attempt to use the backup manual fuel control monitoring system, he said.

Bragagnolo experienced an emergency in the AMX during flight test. Specific test equipment in the aircraft shorted, causing an electrical fire and damage to 700 wires in the aircraft. There was no electrical power, and Bragagnolo had to use the mechanical flight control system. He landed into the barrier at the airport.

As we neared a checkpoint prior to reaching the Alenia base, a circle appeared around the velocity vector in the head-up display while in the navigation mode. Bragagnolo said that we were 2 min. from the next checkpoint. The circle started to diminish as we approached the checkpoint.

The Alenia prototype AMX is equipped with an instrument landing system so that Bragagnolo can land back at Caselle in bad weather. The Italian air force does not have the ILS installed, and at present, their AMXs are equipped with Tacan, along with the Litton INS.

During the flight back to the Alenia facility, Bragagnolo said that in-flight refueling in the AMX is comparatively easy because of the short nose of the aircraft and the uninterrupted clean flow of air around the probe. It is much easier to refuel the AMX in flight than the Panavia Tornado, which he also flies, Bragagnolo said.

I flew the AMX over the active runway at the Caselle airport at 1,000 ft. and 300 kt. The speedbrakes were effective in slowing the aircraft, although they were not needed. They did not produce any pitch movement. Lowering the landing flaps at the 180-deg. position in the pattern did cause a very slight ballooning effect. A tight pattern was flown at 140 kt., decreasing to 130 kt. on final. Bragagnolo had briefed me to fly the AMX almost to the ground prior to flaring and reducing power. He said that once behind the power curve, the speed loss could be 30 kt./sec. if power were reduced too soon.

The angle of attack was 10 deg. during final and increased to 12 deg. at touchdown. The tail would hit the runway at an 18.5-deg. angle of attack, Bragagnolo said. The touchdown was relatively smooth, and a touch-and-go was performed. The head-up-display and the velocity vector were used for touchdown reference for two touch-and-goes and one full-stop landing. The all-around visibility and the 18-deg. over-the-nose visibility allowed for excellent situational awareness in the traffic pattern of the joint military/commercial-use airport. The RB168-807 engine was responsive at the lower end of the power range, although I did not need to make many power changes during the approach and landing. The AMX's response to flight control inputs also was precise and predictable in the landing configuration.

Aviation Week & Space Technology *Managing Editor David M. North flew the AMX from the front seat with Alenia Chief Test Pilot Napoleone Bragagnolo in the rear seat.*

Each landing was a little smoother than the previous. The final touchdown was at 120 kt. By using the antiskid system the stopping distance was kept to 600 meters (1,980 ft.).

By fully depressing the brakes sooner and activating the antiskid system quicker, Bragagnolo said I could have stopped the AMX in 500 meters (1,650 ft.). Embraer is having problems with the antiskid system in its aircraft and will replace a locally built control box within the next few months.

Total flight time was slightly in excess of 1 hr., while the engine start to shutdown was closer to 1.5 hr. The total fuel used for the flight was 4,000 lb. At least half the flight had been flown in the low-level environment.

The AMX partners have developed an aircraft that is a newer technology successor to the A-4 Skyhawk by a wide margin. The AMX handles well in the low-level regime, carries a respectable payload, has a contemporary weapon delivery system, has room for growth and has a price tag in the $13-million range. The cost is fly away and does not include support, training or special requirements. Although some fighter/attack pilots might look down on the single-engine subsonic AMX, it is an honest aircraft for its mission.

Following the flight, I was able to do a more detailed walkaround of a single-seat AMX. All of the access doors were open, and it was evident that the aircraft was designed with ease of maintenance and line servicing in mind. Almost all of the interior units could be reached from standing on the ground. A space in the right side of the fuselage, below the cockpit, has been saved to install a pallet-mounted camera for the reconnaissance mission.

Alenia and its partners also are evaluating other modifications and new weapon systems that could be added to the AMX. Some of these modifications could be adopted by the Italian and Brazilian air forces or for the export market.

Future enhancements

The digital data bus system (Mil-Std1553B) in the AMX provides a flexibility to change the avionics system. A moving map display is being developed and could be installed on the third batch of production aircraft. The map display could be placed on the current multifunction display or a second one to be added later.

A forward-looking infrared system (Flir) has been proposed to the Italian air force for the night-attack role. The Flir would be mounted internally in the forward right side of the nose. The gun is mounted on the left side of the Italian AMX. The Brazilian version has a gun on each side of the nose. The Flir could be added after a laser designator is provided. Alenia has studied the Thomson-CSF Atlis 2 laser designator and the convertible laser designator pod (CLDP). These systems could be incorporated in the 1993-94 timeframe.

Alenia has completed the first phase of weapon integration with Aerospatiale by carrying the Exocet antiship missile on the AMX. The antiship mission is not a requirement for the Italian air force, so Alenia characterizes the Exocet program as a private joint venture. The Kormoran, Harpoon and Sea Eagle missiles also are being evaluated, and Alenia officials said that the NFT Penguin could be integrated into the AMX if there were a demand for the Norwegian missile.

An improved radar is being developed by Brazilian Tecnasa Electronica and SMA-Segnalamento Marittimo ed Aereo of Italy for the Brazilian air force. The proposed SCP-01 is a multimode radar with look-down capability. The radar has been flying on a prototype, and Embraer officials estimate it will be cleared for operational flight within two years.

To date, the AMX is achieving better than predicted reliability and maintainability in operational service, Alenia's Avagnina said. The design goal for the AMX was 12 maintenance man-hours per flight hour. With more than 70 AMXs flying with the Italian air force, the maintenance hours are between eight and nine for every flight hour. The mean time between failures also is tracking less than the design goal of 5 hr., Avagnina said.

SINGLE-SEAT AMX SPECIFICATIONS

Length	13.23 m.	43.42 ft.
Wing Span	9.97 m	32.71 ft.
Height	4.55 m.	14.94 ft.
Wing Area	21 sq. m.	266 sq. ft.
Wing Sweep (25%)	27.5 deg.	27.5 deg.
Empty Weight	6,700 kg.	14,770 lb.
Max. T.O. Weight	13,000 kg.	28,600 lb.
Max. Ext. Weight	3,800 kg.	8,500 lb.
Max. Internal Fuel	2,700 kg.	6,640 lb.
Engine Thrust	5,000 kg.	11,030 lb.

Lavi-TD cockpit reflects pilots' combat experience

David M. North/Ben Gurion International Airport, Israel
March 25, 1991

The high degree of pilot's situational awareness embodied in the Israel Aircraft Industries Lavi Technology Demonstrator reveals what can be accomplished in the design of a strike fighter when combat-experienced flight crews are involved throughout development and testing.

Situational awareness is one of the more important elements in aerial combat, be it air-to-air or in air-to-ground operations. Ultimately, good human engineering is what produces this key feature.

Israel Aircraft Industries was able to use the experience gained by Israeli air force pilots during the many conflicts and wars in the country's more than 40 years of existence. IAI officials said they were able to incorporate lessons learned during the 1982 war in Lebanon, when the Israelis flew McDonnell Douglas F-15s, F-4s and General Dynamics F-16s, among other aircraft, against Syrian Soviet-built aircraft. Using this combat knowledge in all levels of the Lavi's development, IAI has designed a cockpit that ranks at the top of those in modern Western fighters flown by allied forces during the Persian Gulf war.

The cockpit integration, reduced pilot workload and enhanced situational awareness embodied in the Lavi far outclass those features in the Soviet Sukhoi Su-27 and Mikoyan MiG-29, which I flew last year (AW&ST Sept. 24, 1990, p. 32; Feb. 26, 1990, p. 36). However, both Mikoyan and Sukhoi designers are striving to achieve a level of system integration similar to the Lavi's.

Israel canceled the Lavi program in August, 1987, after approximately $1.5 billion of U.S. and Israeli money had been spent on it. The U.S. still has an exchange agreement with Israel for Lavi technology, but as yet there has not been a flow of technical information between the two countries.

The first two Lavi prototypes, which accumulated 82 flights prior to being grounded, are not currently being flown. The third Lavi was scheduled to be the avionics demonstration aircraft in the flight test program. Following the Lavi cancellation, the Israel Defense Ministry funded continued research and development with the third aircraft until IAI picked up the funding in July, 1988. The No. 3 aircraft was transformed into the Lavi Technology Demonstrator (TD) and was flown for the first time on Sept. 25, 1989, by Menahem Shmul, IAI chief test pilot and director of flight operations.

Israel Aircraft Industries has continued to develop the Lavi TD to demonstrate the company's ability to build state-of-the-art high-technology fighters. More importantly for IAI, the company has used the Lavi technology as a showcase for cockpit integration and is offering it as a modernization program for Northrop F-5s and other aircraft. IAI calls the upgrade program the F-5 Plus.

This *Aviation Week & Space Technology* pilot was invited to fly the Lavi-TD in Israel just before the outbreak of the gulf war in mid-January. During the first day, IAI pilots and engineers briefed me on the aircraft's systems. The second day was devoted more to the avionics and cockpit displays, and I had the opportunity to operate the missionized equipment on a test stand.

The Lavi was designed as a multirole aircraft to replace McDonnell Douglas A4s and the Dassault/IAI Kfirs in the Israeli air force inventory. The initial emphasis was on the air-to-ground role, followed by the air-to-air mission. The IAF also needed an aircraft that could be used in advanced and operational training. The service wanted a lightweight aircraft that was relatively easy to fly in training because many Israeli pilots are reserve officers who log minimum hours during peacetime.

On the third day of my visit, I flew the Lavi with Shmul to determine whether the company had met its claims with regard to the TD. Since I do not fly regularly, I was in a good position to find out whether the aircraft was easy to handle and if situational awareness was a strong point.

Demonstrated flight envelope for the Lavi-TD includes a maximum speed of Mach 1.2, an altitude of 40,000 ft. and a maximum angle of attack of 25 deg. Maximum g-forces attained is 7.2.

Strap-in to front seat

The Lavi-TD was manned near midday in an open, two-sided single-aircraft hangar. IAI also has a ground station that monitors performance parameters and actual flight scenes viewed through the head-up display. I took the front seat as briefed, and Shmul helped me strap in. He occupied the rear seat. The aircraft is equipped with a Martin-Baker lightweight Mk.10 ejection seat. It has a 18-deg. aft cant. Shmul said IAI was able to pare the weight of the Mk.10 seat down from 290 to 260 lb. for the Lavi.

We were required to wear both arm and leg restraints in case we had to eject. While the added jacket containing the arm restraints was uncomfortable at first, I did not notice it once we were airborne.

The Lavi's cockpits are nearly identical except for the lack of a head-up display in the rear cockpit. The information on the HUD can be shown on one of the multifunction displays in the rear cockpit. Engine shutoff also has to be done from the front seat.

The Lavi had a maximum weight of 24,370 lb. prior to engine start. The internal fuel load was 5,580 lb., close to the maximum usable internal fuel of 5,785 lb. The Lavi also is capable of carrying three external fuel tanks. The aircraft was equipped with two dummy air-to-air missiles on the wingtip positions, two pylons and a flight-test instrumentation pod on the centerline station. Maximum takeoff weight of the aircraft is 40,500 lb.

The single Pratt & Whitney 1120 engine was motored 40 sec. to even loads on the shaft, and once the throttle was moved to the idle position it required 12 sec. to light off. Both N, high-pressure compressor speed and engine temperature were monitored during engine start.

IAI has lowered the maximum temperature allowed on the 1120 engine from 970C to 960C to extend the life of the three Pratt & Whitney engines in the Lavi TD program. Lowering the maximum limit by 10 deg. doubles the life of the engine, Shmul said. Lowering the limit also reduces thrust in afterburner to approximately 18,600 lb. from 20,600 lb. Military power also is cut to slightly less than 12,500 lb. of thrust from 13,250 lb.

A simplified system is used in the TD to bring the avionics and displays on line in the cockpit. A control box is mounted on the right console with on and off buttons for each of the systems. There also is a button marked training and one labeled combat. By depressing the training button, all of the avionics functions required during a training flight are brought on line at the same time. Depressing the combat button brings the training functions on line as well as other systems, such as weapon and electronic countermeasures. Quick response and very short alert time are essential in a country as small as Israel, which has many potential enemies, Shmul said. We taxied the TD out of the hangar and made a sharp turn using the nosewheel steering button on the stick. The Lavi's nosewheel can pivot up to 35 deg. from the centerline.

We then completed the short pre-takeoff checklist before lining up on Runway 30 at Ben Gurion airport. Once the ejection seat was armed, the "no takeoff warning" on the HUD disappeared.

The wind was from 270 deg. at 15 kt. Temperature at takeoff was 15C, or close to 60F. I advanced the throttle to military power, and the Lavi-TD accelerated quickly to the 140-kt. rotation speed. The takeoff roll was approximately 2,200 ft. with the downrated engine. Shmul said that in afterburner, the takeoff roll would have been about one-half the distance.

I rotated the nose of the aircraft to 10 deg. on the head-up display, and raised the landing gear before reaching 200 kt. After takeoff, Shmul advised me to turn right and follow the steering information given on the HUD and the tactical situation display.

During the flight to the assigned tactical area, I was able to study the cockpit display more closely. The Hughes holographic head-up display provides primary flight information. The HUD has a wide-angle display similar to that of the General Dynamics F-16C. Below the HUD is the up-front control panel. At the top of the instrument panel are three multifunction displays aligned in a row. The color tactical situation display is to the far right, while the other two headdown displays are monochromatic. The avionics package is based on distributed processors interconnected by dual 1553B multiplex buses and controlled by two identical display and control multifunction processors.

Normally, a radar presentation would be shown in the center or left multifunction display, but the Lavi-TD does not yet have a radar installed, although the aircraft is designed for one. IAI plans to install an Elta Electronics EL/M-2032 multimode pulse Doppler radar in the TD this spring and summer. The Elta radar has a look-up/look-down detection capability for the dual fighter/attack roles. The radar currently is undergoing flight tests in the company's Boeing 707 test aircraft.

When the Lavi is down for installation of the radar, IAI plans to integrate GPS and inertial navigation systems for improved positioning accuracy. Later in the year, improved multifunction color displays and display processors employing reduced instruction set computer (RISC) technology will be incorporated into the aircraft.

The influence of pilot input on the design of the Lavi's cockpit instrumentation and controls was evident throughout the flight. En route to the tactical area, Shmul also had me change the four primary flight modes shown on the HUD. By moving a switch on the throttle, I was able to select navigation, intercept, air-to-ground or air combat maneuvering.

IAI has made good use of the HOTAS hands-on-throttle-and-stick aircraft and systems control concept in the Lavi. But Shmul had to remind me several times about the locations of appropriate switches or buttons to perform a specific function. The master mode controller is a hatshaped switch on the throttle. The chaff and flare dispenser switch also is mounted on the throttle.

The flight path, altitude, airspeed, heading and general information, including fuel situation, are presented in the tactical navigation mode on the HUD. The positions of the more than 50 airfields in Israel are contained in the mission computer, and the routes to them from the TD's present position are available on the tactical situation display. The 5-in. color display show-navigation routes in green, while other colors could show borders or tactical lines. Enemy surface-to-air missile sites and enemy aircraft would be displayed in a different color.

The Lavi's mission computers have been programmed for self-defense, with integrated active electronic countermeasures, synthetic voice warning, threat analysis and release of chaff and flares. A quick look at the programming for threat analysis showed a wide range of Soviet aircraft, plus many U.S.-built aircraft. Shmul said that IAI had to incorporate all of the aircraft threat parameters because some of Israel's traditional enemies either are or could be operating U.S.-built aircraft.

Up to this point, I had been flying the TD solely by use of the HUD with quick references to the tactical situation display. There had not been any need for me to take my hands off the throttle or stick.

The weather during the flight consisted mostly of clear skies with cloud buildups over all of Israel. While transitioning from clear to overcast conditions and during some cloud penetration, I found that there was excellent clarity and resolution in the color tactical situation display and other displays. There was no need to change the intensity of the displays.

First ground attack run
We changed the HUD display into the air-to-ground delivery mode once in the tactical area, and selected an abandoned airfield as the target for the first ground attack run. Shmul selected the constant computed impact point mode for the first attack, and CCIP was dis-

played on the head-up display. Roll-in was from 12,600 ft. at 330 kt. The cue for release came at 4,200 ft. and 450 kt. after following the flight path line. On the TD's second air-to-ground test flight, a bull's-eye was scored with the third Mk.82 bomb dropped.

We made the next attack run in the automatic delivery mode. I made the entry with a split-S maneuver from 12,000 ft. The run-in to the designated target was at 3,000 ft. at 400 kt. The HUD symbology showed the route to the target and elapsed time remaining to bomb release. Following the countdown to bomb release, an X-mark indicated a pull-up. The TD's air-to-ground weapon system allows offset bombing as well.

Throughout the flight, the system continually displayed fuel remaining on board above mission requirements.

Shmul had arranged for an Israeli air force F-4 to participate in a demonstration of the Lavi's air-to-air capability and the cockpit displays associated with aerial combat. He warned me that the speedbrakes were not operative, so that I would not rely on them in case we encountered a situation in which we were rapidly overtaking the Phantom. The composite material speedbrakes on the TD are bolted closed. IAI did not have time to evaluate the performance of the speedbrakes in the two prototype Lavis before the full-scale development program was canceled, and Shmul said that it has not been worthwhile qualifying their use on the TD.

The first run on the F-4 was made from 7,400 ft. and 220 kt., while the Phantom was in a racetrack pattern at about 3,000 ft. The encounter was disengaged at 427 kt. at the F-4's six-o'clock position. In the air-to-air mode, the aiming pipper was used and the projected hot line of bullets was shown on the HUD.

Shmul then asked the F4 pilot to perform some tighter air combat maneuvers while we were still in his rear quadrant. I was able to keep the gunsight reticle on or near the F-4 throughout the maneuvers, which Shmul said were designed to make tracking more difficult. The tracking ranged from 200 kt. at 14,000 ft. to 3,000 ft. at close to 500 kt. while pulling up to 4g.

The visibility from the front seat, which was excellent during the entire flight regime, was especially appreciated during the air combat maneuvering and trail exercise with the F-4. The canopy, which is similar to that in the USAF/General Dynamics F-16, is constructed of 5.6-mm. thick stretched acrylic. The F-16 canopy is 11 mm. thick. I noted no distortion in the Lavi canopy.

In several instances, I had to select afterburner power to stay with the maneuvering F-4. The TD's 1120 engine has a five-stage afterburner, but Shmul said the pilot is able to detect only four of the stages in acceleration. I found the acceleration to be very smooth and could not detect the multiple stages. The terms afterburner and military power are displayed on the HUD when those power settings are selected. During the entire flight, the engine's response was smooth. Shmul said that in the more than 150 test flights in the Lavi prototypes and the TD, the pilots "have been able to fly the aircraft and not the engine."

After the F-4 left the area, I did a number of aerobatic and slow-flight maneuvers. My aileron rolls at 300 kt. and 12,000 ft. were uneven at first, but they became crisper as I adjusted to the stick force needed to produce a roll rate of approximately 200 deg./sec. Maximum roll rate of the TD is 300 deg./sec. While the pitch rate of the Lavi is rapid, Shmul said that it is not as good as the Grumman F14's.

I then reduced the power to idle at 14,000 ft. and lowered the landing gear. I flew left and right turns down to 129 kt. with quick reversals, and found no tendency for the nose to wander. I raised the landing gear at 131 kt., and as with gear extension, noted little pitch change.

The Lavi-TD airframe features a closecoupled active canard and a delta wing with no trailing edge flaps. The canard and elevons are controlled by a digital, by-wire flight control system. The aircraft also has relaxed static stability.

At 12,000 ft., I selected afterburner power and raised the nose to a 28-deg. climb angle while maintaining 320 kt. At 16,400 ft., I advanced the throttle to military power to start a loop. I initiated the loop at 380 kt. with a 5.4g pullup. Power was advanced to afterburner in the first 30 deg. of nose-up travel. At the overhead position, the g-force had been relaxed to less than three, at 20,200 ft. and 122 kt. The angle of attack at the overhead was 20 deg. The loop was finished at 16,500 ft. and 10 deg. off the original heading. The Lavi was designed to withstand 9g, but the TD has been cleared to only 7.2g. The design maximum speed is Mach 1.85.

The navigation mode was selected for the return to Ben Gurion airport, and a box on the HUD indicated the airport location. I selected a heading for an instrument landing approach. During the descent from 12,300 ft. at idle power, the airspeed held steady at 230 kt. and the rate of descent was 2,300 fpm. Fuel flow was 1,200 lb./hr. The information for the instrument landing approach was shown on the tactical situation display in both English and Hebrew.

During the entire flight, I felt comfortable flying the aircraft in the head-up mode using only the information on the Hughes HUD. The tactical situation display was used only for reference when needed for navigational bearings. When Shmul reminded me that the ILS bearing and altitude were not shown on the HUD, I had to use the attitude director indicator to fly the approach. It took several minutes for me to go head-down and look at the small ADI, which is located below the center multifunction display.

Primary flight display in the Lavi-TD cockpit (above) is the Hughes holographic head-up display. Below the HUD is the up-front controller. The color multifunction display to the right contains tactical information. The navigation information is in green while tactical information is in other colors. Although the TD is not yet equipped with a radar, a radar display is shown in the mockup's center monochromatic head-down display. While barely visible in the left multifunction display, aircraft weapon loading is presented. Engine instrumentation for the Pratt & Whitney 1120 and fuel quantity are below the left display. The Lavi-TD has a center stick providing input to the digital, quadruple, multimode full authority fly-by-wire flight control systems.

View from the Lavi's Hughes head-up display is just prior to the Israeli air force F-4 being brought under the gunsight aiming reticle during the evaluation flight. The reticle display does not include a radar ring because a radar has not yet been installed in the TD.

I then used the ADI for the initial part of the approach and the HUD information for the final descent. The initial approach was flown at 160 kt. with the landing gear extended. The approach was made with a 16-kt. tailwind. At 150 kt., the angle of attack was 11.6 units. Just before touchdown, the angle of attack was 13 units at 145 kt. I started a slight flare at 85 ft., and the landing was smooth.

Following a touch-and-go, I flew a teardrop pattern to land on the same runway in the opposite direction. During the initial touch-and-go landing and the tight pattern for the second and final landing, I was extremely comfortable with the handling characteristics and performance of the TD. The second landing was smoother than the first. I maintained directional control with the rudder until reaching 60 kt., at which time I applied the brakes to slow the aircraft. I made a 180-deg. turn on the runway at 10 kt., and taxied the TD back to the IAI ramp area.

The block time for flight was 1.2 hr. and the actual flight time was 1 hr. The total fuel consumed was 4,850 lb. for the flight, but nearly 400 lb. had been used at the blocks prior to flight while checking the test equipment.

The cockpit technology used in the TD makes it possible for the pilot to fly in the head-up mode most of the time. I found that my scan was almost entirely out of the cockpit during the flight. The few times that reference had to be made inside the cockpit were because a system had not been integrated with the HUD. The HOTAS concept in the TD has been well defined, and with a little more flight time, I would have become quite comfortable with it. The information presented on the HUD and the tactical situation display has been well thought out and does not overload the pilot with unnecessary details.

The lessons learned from the 1982 conflict with Syria, in which there had been a heavy concentration of surface-to-air missiles, has been included in the design of the TD, Shmul said. "In combat there is a lot of stress, so you have to make quick tactical decisions with one glance at all the relevant information," he said.

The Israel Aircraft Industries goal of developing a cockpit that would have the highest degree of situational awareness has been more than accomplished, as exemplified by the TD. From my experience in flying fighter/attack aircraft, as well as other aircraft, the TD's cockpit technology ranks with the best, if not at the head of the list. The cockpit technology incorporated in the U.S. Air Force's advanced tactical fighter has yet to be evaluated.

IAI also has achieved its second goal in designing an aircraft that pilots with minimal recent flight time would find easy and safe to fly. I was comfortable with the TD in every regime. The fact that I had Menahem Shmul, the Lavi chief test pilot and an outstanding pilot in the back seat, naturally bolstered this feeling of confidence.

Aviation Week editor
flies top Soviet interceptor

David M. North/Farnborough
September 24, 1990

Prior to flying the Sukhoi Su-27, Aviation Week & Space Technology's *Managing Editor David M. North made flights in the
McDonnell Douglas F-15 and the F/A-18 with the U.S. Air Force
and Navy, respectively. The flights were to prepare him to better
evaluate the top Soviet interceptor aircraft. North flew the F-15
with Air Force Lt. Col. Don Kline, commanding officer of the
27th Tactical Fighter Sqdn, at Langley AFB. The F/A-18 was
flown with Navy Lt. Tom Gurney of the Strike Warfare test
squadron at NAS Patuxent River, Md. North was impressed
with the flying skills of both these pilots and the professionalism
displayed by their units. The Su-27 flight from Farnborough
was facilitated by the assistance of Reuben Johnson, a translator and staff analyst in Soviet Studies with General Electric Aircraft Engines.*

The Sukhoi Design Bureau Su-27 and the Mikoyan MiG-29 represent
the most advanced Soviet fighters in operational use, and both of
them exhibit a superb blend of aerodynamic design with high-thrust,
reliable engines.

This *Aviation Week & Space Technology* editor became the first
American pilot to fly the Su-27 earlier this month during a late after-

*Su-27 has an operational maximum sustained g-loading of +9. with full
afterburner power, the Soviet aircraft has better than a 1.0 thrust to
weight ratio at mid takeoff gross weights.*

noon flight from the Farnborough air show flight line. At the same time, I became one of a handful of pilots outside the Soviet Union to have flown both the Su-27 and the MiG-29. I flew the MiG-29 from Kubinka Air Base, USSR, in January with Valery Menitsky, chief Mikoyan Design Bureau test pilot (AW&ST Feb. 26, p. 36).

The path that led to the Su-27 flight was long and circuitous. It started in May, 1989, with interviews with Mikhail Simonov, general designer of the Sukhoi Design Bureau. This interview was followed by another one in January after I completed the flight in the MiG-29. At

The fuel quantity gauge in the Su-27 is below the radar display and to the left of the aircraft test equipment panel. There is no fuel flow indicator in the Su-27 front cockpit. The Sirena 3 passive radar warning system is in the lower right side of the cockpit. The Su-27 angle-of-attack indicator is on the far left of the round instrument in the upper left side of the instrument panel. The g-force indicator is on the right side of the same instrument. Below the landing gear handle in the Su-27 is a VHF radio installed for flying outside the U.S. The ratchet type dials for changing frequencies were difficult to reach while flying. The control buttons for the four-channel analog fly-by-wire flight control system are mounted in front of the throttles in the Su-27. The small latches mounted on the throttles in the Su-27 have to be depressed going into and out of afterburner. The red handle to the inboard side of the left console is for emergency hydraulic braking. The large head-up display mounting in the Su-27 does restrict some of the forward visibility in the Su-27. The Su-27 has infrared search/track sensors and a helmet-mounted sight capability.

that time, Simonov promised a flight in the Su-27 by the end of 1990. He almost was able to keep that promise in Oklahoma City in June. Following a 3-hr. discussion on flying and my aviation background, plus review of appropriate medical certificates, he cleared me to fly the Su-27 during the Oklahoma air show. That flight was canceled by a Soviet bureaucrat just before I entered the Su-27 cockpit with Sukhoi chief test pilot Viktor Pougachev (AW&ST July 16, p. 9). The invitation from Simonov to fly the Su-27 came at the Farnborough air show a week before the beginning of the eight-day event.

Since the beginning of the year, the Soviets have exercised glasnost at its fullest in aviation circles by allowing numerous pilots to fly the MiG-29 with Valery Menitsky, both in the Soviet Union and during air show tours in Canada and the U.S. Only two other non-Soviet pilots had been allowed to fly the Su-27 prior to my Sept. 4 flight, Pougachev said. A Singapore air force general and Ray Funnell, chief of the Australian air force air staff, flew the Su-27 during the Singapore air show in February.

Following my flight, Sir Peter Harding, air chief marshal and chief of the Air Staff of the Royal Air Force, flew the Su-27 with Pougachev on Sept. 7.

Short British ATC form

The day of the flight, Pougachev gave me another checkout in the front seat of the Su-27UB two-seat fighter. Following the cockpit familiarization, we went to the air show control room next to the tower to request permission for the flight. Expecting some bureaucratic opposition over flying a Soviet fighter in British airspace, we were pleased when the only question was whether this was a customer demonstration flight. An affirmative answer was followed by the filling out of a short form listing the pilots and the flight request. The time of the flight was dictated by the air show routine ending at 5:30 p.m.

I was fitted with a g-suit underneath a two-piece flight line outfit. The helmet, oxygen mask and throat microphone were put on in the front cockpit. Pougachev then manned the rear cockpit of the Su-27UB. The attachments for the parachute and seat restraints were similar to those found in other fighters. The flight gear was comfortable, and the seat was relatively hard. The seat would be fine for 1 hr., but could become uncomfortable after a 3-hr. flight.

The Lyulka Engine Design Bureau AL31F engines are started by momentarily pressing a start button on the right console, after placing the throttle into the flight idle position. An external 26-V. power cart was used to start the two engines, but the aircraft is equipped with batteries for internal engine starts. The engine start sequence appeared slow prior to reaching the idle thrust setting of 660 lb. Outside air temperature was close to 75F, slowing the engine start. A small torch-off occurred in the left engine on start, something that is unusual in the Lyulka engines, Pougachev said later. Both engines settled at a 400C reading at idle, and the peak temperature at start was not much greater than that. The engine temperature is measured from behind the turbine, Pougachev said.

Fuel load at the blocks was close to 5 tons, or 10,000 lb. The maximum internal fuel load of the Su-27 is close to 19,000 lb. of fuel, which is considerably more than the standard F-15, but approximately 3,000 lb. less than the F-15C with conformal fuel tanks. The Su-27 is not equipped to carry external fuel tanks on the wing, but the naval version of the fighter is configured to carry one on the centerline station. The standard Su-27 also is not equipped with an in-flight refueling system, but the naval version has a refueling capability similar to the hose-and-drogue system used by the U.S. Navy (AW&ST Feb. 12, p. 28).

Takeoff preparations

Once both engines were at idle power and taxi clearance to Runway 25 was granted by Farnborough tower, the throttles were advanced for taxi. The ability of the Su-27 to get airborne quickly without multiple ground checks is a definite asset for an interceptor. Nose wheel

Sukhoi Design Bureau Su-27 takes off in afterburner. At the 1990 Farnborough air show, both Aviation Week *Managing Editor David M. North and RAF Air Chief Marshal Sir Peter Harding flew the Su-27B from the front seat with chief Sukhoi test pilot Viktor Pougachev in the rear seat.*

steering is through the rudder pedals and is very positive, with the ability to make sharp turns without much effort.

The Su-27 has two separate 4,000-psi. hydraulic systems used for the nose wheel steering and landing gear and flap retraction. The braking system pressure is reduced to approximately 1,400 psi.

I requested clearance from the tower to line up on the runway. A Soviet translator was in the tower to smooth any communication problems I might have with Pougachev during the flight and to help resolve any emergencies. The Soviet pilot has a limited English vocabulary, and I do not speak Russian.

Once cleared to take the runway, I depressed a small trigger on the front of the stick which provides approximately twice the braking power of the standard hydraulic system. This braking system allows more engine power to be applied at a standstill for short takeoffs. The trigger was released after passing into afterburner power, but before maximum afterburner power was reached. The acceleration was rapid, and at 115 kt. I rotated the nose of the aircraft to a 10-deg. climb attitude. The aircraft lifted off at 135 kt. after a takeoff roll of 1,100 ft. and approximately a 10-sec. roll.

The landing gear lever was pulled out and up to raise the gear, and the flap button was pushed to raise the flaps. There is only an up and down button on the flaps, with no intermediate settings. As I attempted to pull the throttles back to climb power, I momentarily forgot that the latches in front of the throttles that must be depressed to enter the afterburner regime also have to be depressed to come out of afterburner power. At first I thought that Pougachev wanted me to stay in afterburner, but a call from the back seat— "Dave, afterburner"—prompted me to lift the latches and reduce power. By this time, I had exceeded the 5,000-ft. altitude given to me by air traffic control and had to perform a sharp inverted roll to acquire the correct altitude. Not helping was the fact that the altitude instrumentation in the Su-27 is in meters and the speed in km./hr., requiring me to make quick mental calculations.

About this time, I was struck by the realization that I was an American flying an unfamiliar Soviet fighter over the English countryside with a Soviet pilot who did not speak English. However, knowing that Viktor Pougachev is one of the top Soviet pilots and that British air traffic control is extremely professional calmed me. I had been switched to London military ATC for the flight to Boscombe Down. The military control cleared us to operate from 4,000 ft. to flight level 150 (15,000 ft.) and kept us under radar surveillance so we would not stray outside the control area.

During the 5-10-min. flight to the control area, I had time to get a better look at the cockpit instrumentation. During the flight in the MiG-29, and now the Su-27, there had been no question of turning on the radar, the infrared sensors or the head-up display. Earlier requests had been turned down by both the Mikoyan and Sukhoi pilots, so I made no request this time to operate the systems.

The front cockpit of the Su-27 is spacious compared with many Western fighters. The size of the cockpit is more on the scale of the F-15 than other aircraft. Visibility is excellent in all directions, and there did not seem to be any distortion in the teardrop-shape canopy. The head-up display and its mounting did restrict forward visibility more than was expected.

Identical flight instruments

Cockpit instrumentation of the Su-27 is very similar to that of the MiG-29. The attitude gyro, horizontal situation indicator, airspeed and altitude indicators, and rate of climb appear to be identical in both Soviet fighters. The weapon control panel was missing from this Su-27, but would normally have been located in the upper left side of the instrument panel. The engine instruments in both aircraft are different, as is the fuel quantity gauge. Pougachev said the Su-27 does not have a fuel flow meter, or gauge, and one could not be seen.

Sukhoi chief test pilot Viktor Pougachev briefs David M. North on the Su-27 cockpit. Soviet translator behind the two is a commercial navigator and flies in the back seat of the Su-27UB.

The Sirena 3 passive radar warning on the lower right side of the instrument panel appears to be the same as the one mounted in the MiG-29. The aircraft test equipment panel installed in the Su-27 also looks identical to that of the MiG-29.

While the aerodynamic performance of the Su-27 is excellent and is comparable to or better than Western fighters, the cockpit is where some shortcomings are evident. The Soviets believe in designing aircraft that are reliable and simple to operate. But the lack of computer and system integration technology places the Soviet pilot at a disadvantage in terms of situation awareness. The Soviets are flying Su-27s and MiG-29s with cathode ray tubes in the cockpit, but when these are unveiled in the coming months, they still will be behind current Western cockpit technology.

Once in the Boscombe Down area, I commenced three successive aileron rolls to the left. The stick was displaced only half the available distance, but the roll rate was still impressive and was approximately 150 deg./sec. Pougachev said the maximum roll rate of the Su-27 is close to 270 deg./sec. The roll rate is higher than that of the F-15C, but it is in the same roll rate regime as the F/A-18. The stick pressure to achieve this roll rate was very light. Pougachev said that this aircraft belonged to another test pilot who was a former Su-26 aerobatic pilot and preferred lighter stick forces. The single-seat Su-27 flown by Pougachev is programmed with a 30% higher stick force, more in line with military aircraft, he said.

Analog fly-by-wire system

The Su-27, unlike the standard MiG-29, is equipped with a four-channel analog fly-by-wire flight control system. Each control system has its own computer and air data source mounted on the fuselage. There are four control lights on the left console and Pougachev said earlier that if any of the lights came on, to depress the appropriate switch.

I flew several split-S maneuvers to obtain a feel for the aircraft. The first two were flown at different airspeeds, and by using military power (maximum power without afterburner) and pulling 4g I was able to resume level flight in less than 3,000 ft.

The final split-S was started at 220 kt. at maximum power. Attempting to keep a constant 24-deg. angle of attack resulted in a loss of 2,300 ft. and an ending speed of close to 220 kt. During the maneuvers, there was no wing fall-off and no buffet.

A half Cuban eight was started at 375 kt. with a 6.5g pull-up in maximum power. The starting altitude was 8,200 ft. and the ending altitude was 12,000 ft. with an airspeed of 185 kt. Again, there was no buffet and aircraft control was positive throughout the maneuvers.

Afterburner power was used sparingly during the flight, although when it was used, there was only a short lag before the acceleration could be felt. The maximum power of the AL-31F engines in afterburner is close to 27,500 lb. of thrust each, giving the Su-27 better than a 1.0 thrust to weight ratio at mid-gross-weight takeoffs. Military power is close to 18,000 lb. of thrust each for the engines. The afterburner in the Lyulka engine is a single stage with metered fuel, giving a steady power increase from the minimum afterburner setting to maximum power. This is unlike the Pratt & Whitney F100 engine in the F-16 and F-15, which incorporates multiple stages in the afterburner range.

The next maneuver I performed, as briefed earlier by Pougachev, was a lowspeed loop. Entry was at 270 kt. and 6,500 ft. Military power was used as I started the loop pulling 4g. The angle of attack was around 15 deg. as the Su-27 topped out at 9,800 ft. and at 118 kt. The downside of the loop ended at a speed of 192 kt. while pulling close to 24-deg. angle of attack on the bottom of the loop.

I had intended to enter a tight 360-deg. turn at 7g in afterburner, but Pougachev indicated that I should stay in military power because of the 426-kt. entering speed. Following the flight, Pougachev said that at 426 kt., we would have exceeded Mach 1 during the turn in afterburner power. General air show limitation restricted us to airspeeds below Mach 1. In military power, the 7g maneuver was performed with the speed the same coming out of the full circle. Below a speed of 370 kt. and entering the turn prior to the application of power, the turn could have been completed without exceeding Mach 1, Pougachev said.

Slow-speed flight was entered at close to 10,000 ft. by first pulling the nose of the aircraft up to a 25-deg. pitch attitude and then using close to maximum power to maintain a 26-deg. angle of attack. There was positive pitch and roll control throughout the slow-speed routine. The speed was 97 kt. and we were descending slightly because of the altitude and the fuel load. Pougachev performs this maneuver during his aerobatic routine and maintains altitude at close to sea level. The 26-deg. angle of attack is a Soviet military limitation, and the aircraft is capable of maintaining a much higher angle of attack.

A similar routine in the Navy F/A-18 flown earlier yielded a 35-deg. angle of attack and a 90-kt. airspeed, with the same amount of controllability. The slow-speed flight in the F/A-18 was at 18,000 ft., and we also had a rate of descent of 2,000 fpm established at military power.

Pougachev then performed the next two maneuvers while I followed him on the stick. At 7,000 ft. and with an entering speed of 320 kt., he pulled the nose up to a 75-deg. pitch angle. The power was set at 85% and at 10,500 ft. the airspeed indicator read zero. The an-gle-of-attack indicator was pegged at its mechanical limit of 38-deg. The Su-27 then slid back for about 3 sec. before the nose fell slowly through the horizontal to a nose-low attitude and Pougachev added power. The altitude at recovery was 8,200 ft.

The Pougachev Cobra was the next maneuver performed by the Soviet pilot. Pougachev had not instructed me how to turn off the angle-of-attack limiter in the front cockpit, so he had to pull an ad-ditional 34 lb. of back pressure to go past the 26-deg. angle-of-at-tack limit. The Cobra maneuver was entered at 220 kt. at 10,000 ft. Pougachev pulled very quickly on the stick and achieved close to a 90 deg. pitch attitude with an 85% engine power setting. The one speed that registered as I followed the maneuver was 83 kt. after the maximum pitch and before power was added to resume horizontal flight. The forward center of gravity of the two-seat Su-27 precludes pitching beyond 90 deg., Pougachev said. During one Cobra routine at the Farnborough air show, Pougachev achieved a 130-deg. pitch attitude.

The low-speed maneuvering capability of the aircraft was im-pressive and the ability to point the nose of the aircraft throughout the slow-speed regime was equally impressive. The large differential horizontal stabilizers provided positive control in slow fight.

During the tail slide and the Cobra maneuver, the two Lyulka en-gines did not stall or give any indication of a problem. The engine temperature appeared to be stable during the maneuvers and re-mained at approximately the temperature corresponding to the 85% power setting.

Although we did not perform any acceleration checks in the Su-27, Pougachev estimated that the time to accelerate from 330 kt. to 540 kt. in maximum afterburner would be less than 10 sec. I found during the flight that power response was quick and positive.

Commenting later on the Su-27, Patrick Henry, McDonnell Dou-glas vice president of flight operations, said that the overall perfor-mance figures and airspeeds of the Su-27 were similar to those of a comparably equipped F-15. For many years, Henry was the F-15 air show demonstration pilot. The agility and lowspeed maneuverability of the Su-27 represented by the Cobra and other maneuvers were truly impressive, he said.

The speedbrake in the Su-27 was not tried in flight, but Pougachev said that the extension limit for the large speedbrake mounted on the top of the fuselage was Mach 0.9. There is no such limit for the F-15, but the F-15's speedbrake will not fully extend until airspeed de-creases below Mach 1.3 because of the air loads, Henry said.

Another system not used during the flight was the panic system. The Su-27—as is the MiG-29—is equipped with a button on the control stick that, once depressed, brings the aircraft to wings level, right side up no matter what the attitude of the aircraft was when the button was pushed. The panic feature is tied into the autopilot system.

The incorporation of this system prompted me to ask Pougachev after the flight about the spin characteristics of the Su-27. At first, Pougachev said that the angle-of-attack limiter prevented operational pilots from departing from controlled flight and putting the Su-27 into a spin. He later added that during flight testing, the Su-27 was put into steep and flat spins, but that the spins had a slow rotation rate and were easy to control.

Following the Cobra maneuver, I once again took control of the Su-27 and asked London military control for a steer back to the Farnborough airport. The clear weather in the area allowed me to see the airport from more than 20 mi. away. Once switched to Farnborough tower, I requested a pass down the runway at 1,000 ft., breaking left to a downwind leg. The break was entered at 300 kt. and in attempting to make a sharp break, I over-controlled to a 110-deg. bank angle, which I quickly corrected. The aircraft was slow to drop below the 250-270-kt. limit for lowering the landing gear and flaps, without using the speedbrake.

Factors affecting touchdown

A fairly tight pattern was flown from the abeam position, arriving at 140 kt. in the short straightaway with an angle of attack of 10 deg. Touchdown was at 125 kt. with only a slight flare and at a 12-deg. angle of attack. The actual touchdown was sooner than I expected, primarily because of the tall landing gear and the high attitude on approach. The aircraft was stable during the approach and landing and responded easily to small pitch and power changes.

I did not fully depress the brakes to take full advantage of the automatic braking system, so landing roll was slightly long at 1,900 ft., rather than the 1,650 ft. normally used during maximum braking. The Su-27 was taxied back to the fighter line at the Farnborough airport, just as the sun was setting.

Total time from blocks to blocks was 1.1 hr. and the fuel used was 7,400 lb. All of the flight time had been below 15,000 ft. and a small amount of the total in afterburner. The Sukhoi Design Bureau is proud of the large internal fuel capacity of the Su-27 and the relatively efficient Lyulka engines, which provide a range with maximum fuel of more than 2,000 naut. mi.

My overall impression of the Su-27 was of a large fighter, with the agility and quickness of controls more representative of a smaller aircraft. Slow-speed characteristics are excellent, and a good Soviet pilot would be able to point the nose of his aircraft and fire a missile within a large combat envelope. I felt that I had full authority over the aircraft and was comfortable with its performance and handling characteristics. The Su-27 is more comparable to the improved performance F-15E than to the earlier model F-15s.

Another impression was that Viktor Pougachev must rank as one of the top Soviet pilots. He was easy to fly with and any request to do something with the aircraft was met with a "no problem" response. He had a deep understanding of the capabilities of his aircraft, as well as the limitations. If he and Mikoyan chief pilot Valery Menitsky are a true representation of Soviet military pilots, they would be formidable opponents in the sky. Under glasnost, they both have become friends and compatriots to a growing number of Western pilots.

Aviation Week editor flies Soviet-based MiG-29 fighter

David M. North/Kubinka Air Base, USSR
February 26, 1990

The Mikoyan Design Bureau MiG-29 is an extremely agile aircraft at slow speeds, reflects the Soviet design philosophy of simplicity and reliability and, in the hands of a skilled pilot, is a lethal weapon against comparable Western-built fighters.

In the spirit of glasnost, the Soviets accepted my request to fly their latest Mikoyan fighter from the Kubinka air force base near Moscow on Jan. 30. While the flight itself was important because it provided a first-hand look at the Soviet fighter, the fact that it took place from an operational MiG-29 base represented a big step for glasnost.

On that cold winter afternoon at Kubinka, this *Aviation Week & Space Technology* pilot became the first journalist to fly the MiG-29. During the after-flight dinner in the Kubinka reception area, U.S. and Soviet guests said this gathering would not have been possible a year ago. Hosts Col. Vladimir Sokolov, Kubinka base commander, and Maj. Gen. Nickolai Antoshkin, commander of the air force in the Moscow region, agreed.

Evaluation flight filmed

As a part of this openness, the Soviets also allowed an *Aviation Week & Space Technology* video team to film the MiG-29 evaluation flight and a demonstration of the Sukhoi Design Bureau Su-27 the following day. Sokolov said many delegations have visited the Kubinka facility, the showcase air base for the air force, but a video team had never before been given such extensive leeway. However, during the two days at Kubinka, there were no MiG-29 operational flights, partially due to the weather, and we were not allowed to photograph the MiG-29 air wing at the other end of the main runway in its revetments.

The Kubinka air base air wing was the first to receive operational MiG-29s in 1983. The prototype MiG-29 first flew in 1977. The first U.S. citizen to fly the MiG-29, Benjamin Lambeth, director of the international security and defense policy program at the Rand Corp., also flew from Kubinka in December, 1989 (AW&ST Jan. 1, p. 35). Lambeth had become friends with Valery Menitsky, Mikoyan's chief test pilot during air shows and visits by the Soviets to the U.S. Lambeth speaks Russian and is an expert on Soviet military policies, as well as a pilot with numerous hours in contemporary U.S. fighters.

Maj. Robert Wade, a Royal Canadian Air Force CF/A-18 pilot, became the first North American to fly the MiG-29 at the Abbotsford air show in August, 1989. Wade is the only Westerner to have flown the Soviet fighter from the front seat. He does exhibition flights in the McDonnell Douglas CF/A-18 and was flying at the air show in British Columbia.

While the Soviets are becoming much more lenient in allowing access to their aircraft at air shows and in authorizing MiG-29 flights, they still will restrict certain maneuvers and use of aircraft equipment. When I asked Menitsky to fly more of an operational profile than an aerobatic one, he said we should investigate the agility of the aircraft at low altitude and slow speeds, which is one of the MiG-29's strong points.

I also said I would like to climb to a high altitude and accelerate to close to the aircraft's maximum advertised speed of Mach 2 plus to reflect a high-altitude intercept. Menitsky said we did not need to go fast and that being close to Moscow, it was difficult to obtain clearance for a high-speed run. Wade had asked Menitsky during his August flight if he could use the head-up display mounted in the front cockpit for reference, but was told it was not necessary for the flight.

The *Aviation Week* contingent arrived at the Kubinka air base at 10:15 a.m. on Jan. 30 for a possible flight before noon. The overcast was ragged at close to 200 ft. and the visibility was less than a half-mile in light snow at our arrival. The Soviet ground crews were working to clear the runway and taxiway of snow and ice with the exhaust from MiG-17 turbine engines mounted on trucks.

Design of the MiG-29 fighter reflects the Soviet desire for reliability and simplicity. One of the RD-33 engines can be replaced and tested by five technicians in less than 1 hr.

While waiting for the weather to clear, I requested a cockpit checkout in a single-seat MiG-29 sitting in the hangar behind the reception building. The base commander and the MiG representative in the group, Ushkintsev Konstantin Alexandrovich, authorized Lt. Col. Mihail Chadin, senior engineer of the Kubinka division, to conduct the briefing.

The first impression of the MiG-29 cockpit is that it is comparable to the late-1950s McDonnell Douglas F-4 aircraft with analog instruments and round dials. The MiG-29's instrument panel does not contain the digital and cathode-ray-tube-based displays found in many current Western fighters, such as the F/A-18 and F-15. The instruments and displays suggest that the Soviets do not pay as much attention to situational awareness for the pilot as do Western designers.

However, the cockpit design and the entire MiG-29 reflect the Soviet operational philosophy that for a weapon to be effective it must be simple, reliable, economical to manufacture and easy to repair.

"The MiG-29 has been designed for a high combat readiness and to operate from rough fields in any climate," Rostislav Belyakov, general designer for the Mikoyan bureau, said during a preflight briefing. "It has a 1.5 times better mean-time-between-failure rate than that of the F/A-18 and it is getting better."

The Mikoyan bureau has been flying a MiG-29 with cathode ray tube displays and a fly-by-wire flight control system for two years. The decision whether to put all of these refinements into the next batch of production aircraft will be based on whether the modifications meet the Soviets' reliability standards.

While the Soviets strive for reliability and simplicity in their weapon systems, the fact is that because of their lagging technology base, it would have been difficult for Soviet design bureaus to incorporate all the electronic advances of an F/A-18 when the MiG-29 was developed in the late 1970s.

Flight controls compared
The MiG-29 does contain many systems from earlier fighters, but it is a further departure from its ancestry than earlier MiGs. The proof that the Soviets do the best with what they have was established during the flight. The low-speed performance of the MiG-29 with its wire-and-pulley and servo-hydraulic flight control system was comparable to that of the F/A-18 with its fly-by-wire control system.

The visibility from the front seat of the single-seat MiG-29 is excellent. Sitting in the cockpit, I could look down almost the entire top section of the aircraft's nose. I could see a large portion of the wings, sections of the vertical stabilizers and even the AA-10 Alamo medium-range and AA-8 Aphid infrared air-to-air missiles mounted on launchers on the wing's hard points.

The canopy bow also contains numerous mirrors to observe the six-o'clock position. The forward extension of the vertical stabilizers also could be seen. Chadin said that after the experiences in Afghanistan, the flare and chaff dispensers were mounted here so they would be ejected from the top of the fuselage.

The static display MiG-29 was not powered, but Chadin said the MiG-29's head-up display (HUD) portrayed the standard flight symbols, such as roll, pitch, heading, speed and altitude. The information from the infrared search and track system mounted just forward of the windscreen can be displayed on the HUD. The fighter also is equipped with a laser rangefinder. The single-seat MiG-29 is equipped with a multimode pulse Doppler radar with the NATO code name of Slot Back. The radar display is about 3-4 in. wide and mounted on the upper right side of the instrument panel. The radar display also can be shown on the HUD.

On both sides of the HUD base were rectangular windows and prisms of glass that are used in conjunction with the pilot's helmet-mounted sight. The sight is used for off-axis missile aiming, a feature just now being evaluated for use in Western-built fighters.

The weather lifted to about a 500-ft. ceiling in the late afternoon and by that time the runways and taxiways at Kubinka had been cleared of ice and snow sufficiently so permission was given for the flight. The visibility below the overcast had improved to at least 2 mi. Menitsky launched from the Zhukovsky flight test station and arrived at Kubinka at close to 3:30 p.m.

I was provided with all the required flight gear by the Soviet air force, except for gloves I had brought. The g-suit was worn under a bib-type flight suit. The suit was lightweight and fit perfectly. The g-suit hose fit through a hole in the left side of the flight suit. I wore a lightweight flight jacket over the flight suit, and over that a winter-weight flight jacket with a fur collar. The flight boots had laces in the front, but also zippers in the side for quick donning.

The helmet is worn with a light cloth liner. The white helmet contains a visor that was not used during the flight. The Soviet helmet appeared to be larger, but also lighter, than the U.S. Navy helmets I have used over the years. It fit perfectly, and during the flight I noticed very little outside noise.

Front-seat-only operations

During the earlier briefing, Menitsky had said I would occupy the front seat while he flew from the rear seat. The front and rear cockpits are nearly identical in the MiG-29UB. The two Tumansky RD-33 turbofan engines—rated at 18,300-lb. thrust each in afterburner—can be started from the front seat, but not the rear. The control for increased nosewheel steering is in the front cockpit only.

Because of the weather, Menitsky said he would feel more comfortable flying the aircraft from the front seat. Although disappointed, I agreed with his decision. I would have made the same decision in his place, with an unknown pilot who does not have much recent high-performance fighter time. The fact that I do not speak Russian and Menitsky does not speak much English added weight to the decision. Rand's Lambeth flew rear seat, and poor weather figured in that decision as well.

Because lengthy verbal communications would not be possible in flight, Menitsky briefed me the day before. I was to make the takeoff and then once near 8,000 ft. we would do slow flight and stalls. Tail slides were to follow with some low-speed reversals and loops. Several tight turns with high g-force loadings were to be flown and then a touch-and-go landing and a final landing.

Instrument panel of the Mikoyan Design Bureau operational MiG-29 closely resembles U.S. and NATO aircraft developed in the 1960s, although Mikoyan has been flying a cathode-ray-tube-equipped MiG-29 for the past two years. Mounted on the center stick are control switches for target acquisition, autopilot disconnect, identification-friend-or-foe, gun trigger, standard trim and what the Soviets term the "panic button." No matter what the aircraft's attitude, depressing the panic button brings the aircraft wings level and upright.

The Soviet ground crew strapped me into the rear seat of the MiG-29. The ejection seat is inclined nearly 10 deg., closer to the F/A-18 than the approximately 30-deg. angle in the F-16. The parachute and seat strap harness attached at the chest in a common fitting. There are no leg straps as in Martin-Baker ejection seats, but instead restraining cables cinch the legs close to the seat upon ejection.

Menitsky clearly had great confidence in his MiG-29, because the emergency briefing was not extensive. I knew enough to get out of the seat on the ground and pull the between-leg ejection handles, but that was about it. Menitsky said that if communications failed, three quick movements of the stick would indicate an immediate return for landing.

I had confidence in the K-36 seat after watching videotapes of Anatoly Kvotchur's ejection from the MiG-29 at last year's Paris air show. Some Western officials contend that a successful ejection could not have been made from a number of NATO fighters under the same extreme conditions that Kvotchur faced. The oxygen, g-suit and communication connections were on the left side of the cockpit, as in U.S. fighters. The oxygen mask fit snugly, but was not uncomfortable. A throat microphone was used. Lambeth said there also was a microphone in the oxygen mask, but I did not notice it.

Restricted rear-seat view

Once fully strapped into the fighter's back seat, I had the opportunity to evaluate my surroundings while Menitsky strapped in and taxied to the 8,000-ft. Runway 5 at Kubinka.

The MiG-29 was designed as a single-seat fighter, and the position of its rear seat was not optimized to allow an unrestricted view of the front and rear quadrants. I was unable even to see Menitsky's helmet over the rear cockpit's instrument panel. The aircraft's wings were visible only if I stretched forward in the seat.

Mikoyan installed a periscope above the rear seat instrument panel in the canopy center bow. The field of view through the periscope is over the nose of the aircraft. Lambeth said that during his flight, the periscope image was focused at infinity, with the horizon on the mirror precisely aligned with the true horizon. This gave the effect of looking through the ejection seat headrest in the front cockpit. The periscope closes automatically when the gear is retracted and pops up when the gear is extended.

For some reason, the periscope was not operative when I flew the aircraft. One MiG-29 pilot said that while the periscope can be used to make a landing, it should have a larger field of view. Similar periscopes have been installed in earlier two-seat MiGs.

The automatic engine start sequence in the MiG-29 is similar to that of other modern fighters. Menitsky started the right engine first and then the left with the aid of external power. The fighter has an auxiliary power unit, but it was not used. The exhaust gas temperature peaked at about 750C and then settled to 450C. The Canadian pilot, Wade, said the exhaust gas temperature and rpm range of the RD-33 engines were almost identical to those of the General Electric F404 engines in the F/A-18.

Gross weight of the MiG-29 at engine startup was 15,000 kg., or slightly over 33,000 lb. This is the normal takeoff weight for the single-seat MiG-29. Maximum takeoff weight for the fighter is 18,000 kg., or 39,000 lb. Fuel load was 3,200 kg. (7,050 lb.).

Effective FOD doors
Once the engines had been started, the two large inlet doors closed and engine air was brought through the louvers on each wing root leading edge extension. The alternate doors were installed because of possible foreign object damage to the engines, especially at unimproved landing sites. Menitsky said the FOD doors had proved to be very effective.

Menitsky taxied the MiG-29 to Runway 5 and used the alternate nosewheel steering command on the stick. In the normal mode, the nosewheel travels 8 deg. either side of the centerline, and this is increased to 30 deg. by use of the nosewheel button.

Once the MiG-29 was centered on the runway, Menitsky advanced the sidemounted throttles past military power to the afterburner and said "Dave fly." The nose strut had depressed going through military power without the brakes allowing the main wheels to slide. The brakes were released to coincide with afterburner initiation.

The acceleration was instantaneous. Brakes were used for steering control until reaching 100 km./hr. (54 kt.) and then rudder input was used. It required some right rudder to keep the nose aligned with the runway during the take-off roll. At 200 km./hr. (108 kt.), I pulled the stick approximately one third back and the aircraft broke ground at 230 km./hr. (124 kt.). During this time, the front engine inlet doors opened and the top mounted inlets closed. I did not hear the doors cycling, but I did hear them cycle in flight during some of the high-angle flight routines. The takeoff roll was less than 350 meters, or 1,100 ft.

We entered the overcast at approximately 170 meters (550 ft.), by which time a 45-deg. attitude had been established and the landing gear and flaps had been raised by Menitsky. The MiG-29 came out of the overcast at 1,400 meters (4,600 ft.) at approximately 550 km./hr.

(296 kt.). During the climb, some left rudder was required to keep the aircraft from yawing.

Shortly after breaking out of the over-cast, I performed an aileron roll while still accelerating. The aircraft was responsive, but it required more stick input than I have used in other jet aircraft flown recently. Once the input was made, the roll rate was 250-300 deg./sec. The stick is higher and is mounted farther forward than in other aircraft. Wade said the stick was 2 in. farther forward and 2 in. higher than that in the F/A-18.

The same amount of stick travel did not appear to be needed to change pitch in the MiG-29. Throughout the flight, pitch control was responsive without being overly sensitive. Wade estimated that the pitch rate was 30 deg./sec. with 3 sec. required to become vertical. Overall control of the aircraft was excellent, and the Mikoyan bureau has achieved near fly-by-wire handling qualities with the hydraulic servo-operated controls.

Wade and Lambeth performed three consecutive aileron rolls during their flights with Menitsky. Both said they noticed no roll coupling tendencies during these maneuvers.

I slowed the aircraft at idle power and maintained altitude at 1,600 meters (5,300 ft.). The angle-of-attack indicator is located above the glare shield on the left side of the cockpit. As the indicator approached 30 units, there was a slight airframe buffet. The indicator does not measure beyond 30 units, but at an estimated 35 units the stick was full aft and the speed was close to 155 km./hr. (83 kt.). Rudder and aileron control was available throughout the stall entry and was used to keep the nose at the desired position. At the 83-kt. mark, the aircraft's nose fell without any strong tendency to drop off on a wing.

Menitsky had said that a tail slide would be the next maneuver. The nose of the MiG-29 was pulled up to a 75-deg. attitude at idle power. The excess speed on entry allowed a climb of approximately 800 meters (2,600 ft.) until the aircraft achieved a zero speed. Aileron control appeared effective during the entire maneuver.

At that time, the aircraft slid back approximately 150 meters, or slightly less than 500 ft. I had been slow to apply afterburner power, so the slide back had been greater than most performed by Menitsky at air shows. Afterburner power is achieved by lifting toggles on the throttles to go from military power to afterburner. The forward push through a restraining detent to afterburner power, such as installed in the F/A-18, would have been welcome. The afterburners lit simultaneously, and there was more than adequate elevator control to push the nose over and fly out of the maneuver.

During his flight at Abbotsford, Wade had taken the MiG-29 to a full 90-deg. climb, and observed the same responsiveness of the flight controls and the engines. He also found that on the recovery he was able to stop the nose at the 45-deg. position and maneuver the aircraft. The MiG-29's more than 1.2 thrust-to-weight ratio during these flights allows the pilot to point the nose of the aircraft precisely. The responsiveness and reliability of the Tumansky engines also give a pilot the confidence to attempt these maneuvers.

Menitsky said the engine failure experienced by Kvotchur at the Paris air show during his routine was unique. He said chemical analysis of the engines removed from the crashed MiG-29 indicated that a bird strike had occurred before the tail slide routine. He said that during the MiG-29's flight test program, only one flameout was experienced, and that was at 14,000 meters (46,000 ft.) during maneuvers. He also said the last modification was made to the RD-33 engine more than three years ago, and none have been made since the Paris crash in 1989.

At this point, Menitsky said "Valery fly," and executed a 180-deg. reversal. He initiated the maneuver by pulling the nose up to 65 deg. and then adding military power. At approximately 200 km./hr. (108 kt.) indicated airspeed, left rudder was fed in and the nose started to transcribe an arc. Some opposite aileron had to be used to keep the top wing in position. The fighter topped out at 100 km./hr. (54 kt.) and then came down the other side in a reversal. My maneuver was not nearly as smooth as Menitsky's. The exit speed was close to 550 km./hr. (297 kt.).

I then performed a loop in military power, entering at 500 km./hr. (270 kt.). The 4g pull was started and Menitsky said to keep the angle-of-attack units at 13. Loops in afterburner power are initiated at 280 km./hr. (151 kt.). I topped out after going through approximately 500 meters (1,600 ft.) and at 180 km./hr. (97 kt.).

Aircraft attitude system

Throughout these maneuvers, I was having difficulty keeping my situational awareness from the back seat. I was not able to see the nose of the aircraft, nor the wings, so I relied on the attitude reference system. I discovered that unlike U.S. attitude systems, the small aircraft in the middle of the indicator banked in relation to the background. During the loops, the attitude reference ball tumbled, so I had to wait until it settled to pick up an attitude reference.

Menitsky said the newer operational MiG-29s had a better attitude reference system. The requirement to maintain a scan on the different

positions of the primary instruments in the MiG-29 also did not help
my situation awareness.

Having a military attack pilot background, I wanted to do a typical roll in to a target. Picking a dark spot in the overcast, I rolled the
MiG-29 inverted and pulled through, establishing a 45-deg. dive. The
power setting was below military at the time of the roll in. The aircraft
tracked smoothly during the dive, without rudder input. The maximum speed at the bottom of the dive was 800 km./hr. (430 kt.), the
highest speed attained during the flight.

Aviation Week & Space Technology *Managing Editor David M. North is
strapped into the rear seat of the MiG-29UB at Kubinka air base.*

The final flight maneuver was a 7.5g level turn in afterburner. Menitsky called "Valery fly," and he pulled 7.5g starting at 680 km./hr. (367 kt.) and ending the 180-deg. turn at close to 760 km./hr. (410 kt.). After "Dave fly," the same maneuver was done with similar results, but a bit rougher. The maximum sustained g-load on the MiG-29 is 9g, but several MiG-29 pilots at Kubinka said that, if required during air combat maneuvering, they could pull up to 10.5g. During high-g maneuvers, I found the g-suit to be effective.

During Wade's flight in the MiG-29, in August, 1989, he started a 9g turn at 800 km./hr. in afterburner and executed a 360-deg. turn while accelerating. The maneuver was flown at 400 ft.

There was no opportunity to time any of the speed accelerations in flight. However, Menitsky said that in afterburner power it normally takes 8 sec. to accelerate from 600 km./hr. (322 kt.) to 1,100 km./hr. (591 kt.) at low altitude.

I had control of the aircraft as we descended back to Kubinka. We had remained within 30 km. of the field during the entire flight as indicated by the distance measuring equipment displayed in the cockpit. Once in the clouds, I lost attitude reference once and, with a more urgent request, asked "Valery fly" until I could regain my scan and reference. I was tempted to see whether the "panic button" on the right top side of the stick worked as advertised. Earlier in the day, a MiG-29 pilot had told me that once the button was depressed a mode of the autopilot system would immediately bring the aircraft wings level and right side up.

Another mode of the autopilot can be set to bring an aircraft level at a pre-determined altitude, thereby establishing an altitude floor for the aircraft. One mode of the autopilot also sets up for a second approach after a missed approach, the MiG-29 pilot said.

We were to make an instrument landing system approach to Runway 23 at Kubinka air base. The autopilot was not used during this time. Menitsky said that an automatic approach can be made down to about 50 ft., but the aircraft system does not include autothrottles.

The MiG-29 was very stable during the flight to line up for the ILS. Altitude and heading corrections were accomplished with small control inputs. The deployment of the speed brakes mounted between the engines to slow the aircraft to the landing pattern speed of 500 km./hr. did not change the pitch of the aircraft, and could hardly be noticed at the relatively slow speed. Once lined up on the ILS, the landing gear was dropped at 500 km./hr. (270 kt.) and the aircraft slowed to a 290-km./hr. (156-kt.) approach speed.

The ILS indicator registered heading and altitude information, and corrections were made by flying into the command bars. The location

Single-seat Mikoyan Design Bureau MiG-29 is exhibited below a poster of Lenin at the Kubinka air base. The MiG-29 is shown carrying AA-10 Alamo radar-guided and AA-8 Aphid infrared missiles.

of the indicator was low on the left side of the instrument panel, and was not in an ideal location to make low approaches from the back seat. The small scale of the indicator also was not ideal for instrument flying. This would be nullified in the single-seat aircraft if the ILS cues were shown on the head-up display.

I flew the aircraft to the air base perimeter and then asked Valery to fly, because I could not see the runway. He made a perfect landing, with little flare, at a 230-km./hr. (124-kt.) touchdown speed. He made the touch-and-go and then we pulled back up to 2,000 ft. for a final landing. I again flew the pattern and initial approach, with Menitsky making the final landing.

The landing roll was less than 600 meters (1,970 ft.) and the anti-skid system appeared to function well on the rollout. The MiG-29 has air brakes installed on the main landing gear wheels, and the noise could be heard in the cockpit. The control for the air brakes is a handle on the forward side of the stick. The drag chute also was deployed during the landing. Menitsky said that the Mikoyan bureau is looking at ways to improve the braking system.

The total block-to-block time was 50 min., of which 40 min. was flight time. Total fuel used during flight was 1,800 kg. (4,000 lb.) and 300 kg. (650 lb.) was consumed during engine startup and taxiing to the runway. I had been unable to decipher the fuel flow meters during the flight, and fuel flow in afterburner power did not seem to register. Menitsky said that the 1,100 kg. (2,450 lb.) of fuel remaining in the MiG-29 was enough to get him back to the Zhukovsky flight test station, near Moscow.

Wade, Lambeth and I all had high praise for Menitsky's piloting skills and professionalism during our flights. He is a pilot's pilot and flying with him was sheer pleasure. He is considered one of the best pilots in the Soviet Union.

The overall impression of the MiG-29, as flown by Wade, Lambeth and myself, is of a highly maneuverable fighter. As F/A-18 pilot Wade points out, "The Western pilot would be wise to detect and shoot at the MiG-29 from a distance using his high-technology weapon system, because if it comes down to a close encounter with infrared missiles or guns, a good Soviet pilot is a definite threat."

Improved VTOL performance of TAV-8B adds realism to attack-force training

Brendan M. Greeley, Jr./Quantico, Va.
August 3, 1987

U.S. Marine Corps/McDonnell Douglas-British Aerospace two-seat TAV-8B Harrier 2 aircraft now entering service have the performance required to conduct realistic training for pilots who will man the single-seat squadrons in the Marines' all-Harrier light attack force.

Until the arrival of the first TAV-8B last month, Marine Attack Training Squadron (VMAT)-203 at MCAS Cherry Pt., N.C., had been using Hawker Siddeley TAV-8As to train pilots in vertical takeoff and landing operations before soloing in the AV-8B.

The vertical performance of the TAV-8As is severely reduced by the high-density altitudes that prevail throughout the summer at Cherry Pt. Compounding these performance limitations is the small number of aircraft available—the squadron has only seven TAV-8As.

Vertical takeoff and landing (VTOL) operation is not an end in itself, but remains the key to the service's concept of employing the AV-8B. Attaining proficiency in this flight regime is broadly analogous to the requirement that Navy aviators become proficient at carrier operations. Many of the Marine Corps' AV-8A pilots have converted to the AV-8B, but some are still going through the conversion process along with aviators transitioning from other aircraft—primarily the USMC/McDonnell Douglas A-4M—and newly designated naval aviators arriving from the Naval Air Training Command.

McDonnell Douglas-British Aerospace TAV-8B Harrier 2.

82

Pilots who have never flown the Harrier are assigned 15 TAV-8A sorties before soloing in the AV-8B, which means that TAV-8A availability is critical, especially during the summer. The improved handling qualities of the AV-8B may allow a reduction in the two-seat training syllabus to seven or eight sorties, Lt. Col. Ben Mayer, VMAT-203 commanding officer, said. Pilots converting from the AV-8A get a reduced number of sorties based on individual requirements.

This *Aviation Week & Space Technology* pilot recently flew the first TAV-8B, the full-scale engineering development aircraft, from the 4,200-ft. runway here at Quantico. This particular aircraft had 900 lb. of instrumentation on board that increased its empty operating weight to 15,300 lb. The empty weight for production TAV-8Bs is 14,223 lb.; single seaters weigh 13,700 lb. empty.

Accompanying me was Jackie C. Jackson, McDonnell Douglas chief production test pilot, who flew AV-8As while on active duty with the Marine Corps and now flies A-4Ms in the Reserves. Jackson was in the front seat for the 1-hr. flight while I occupied the raised back seat, which normally will be occupied by the instructor.

Visibility is excellent in all directions from the back seat, a key consideration when instructing in any aircraft but particularly so in the Harrier, in which keeping lateral ground reference points lined up is essential to maintaining a stable position when hovering. By leaning slightly to one side, I found it easy to see forward over the nose, also an important consideration during air-to-ground weapons training.

Jackson did the preflight while Nathaniel J. (Jerry) Johnson from McDonnell Douglas helped me strap into the rear cockpit. The aircraft is equipped with Stencel ejection seats with a zero-airspeed/zero-altitude capability and has a command ejection system that allows either pilot to eject the other seat, if required. Both the TAV-8B and the single-seat AV-8B aircraft are equipped with a Clifton Precision, Inc., on-board oxygen generating system that precludes any requirement to service the aircraft with liquid oxygen.

After I had hooked up to the oxygen and communications system, Jackson gave me a radio check on the intercommunications set and then cranked the engine using the on-board self-starter, a Lucas gas-turbine starter activated by the aircraft battery. Simultaneously, he began aligning the Litton AN/ASN-130A inertial navigation system, which took about 4 min. This delay can be eliminated when standing alert by starting the aircraft, aligning the system, storing the update and then shutting down.

The Marine Corps concept of employment for the Harrier, in which aircraft are deployed to austere forward sites and placed on ground alert, dictated the installation of an auxiliary power unit (APU) to keep the aircraft battery recharged. During ground alert with no a.c. power available, the "alert" position of the battery switch routes power to both radios, the secure voice system, utility and kneeboard lights and a d.c. voltmeter. When battery voltage drops, the 6-kva. APU can be started to recharge the battery.

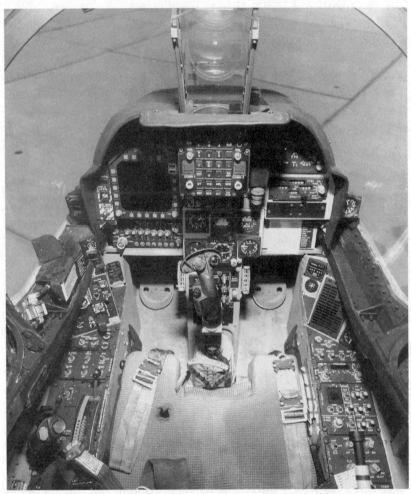

USMC/McDonnell Douglas-British Aerospace TAV-8B front cockpit has coordinate data control panel centered below the Smiths Industries head-up display—the primary flight reference instrument. Stand-by flight instruments can be seen below the data control panel. Large cathode ray tube digital display indicator is left of the data control panel; immediately below the DDI is the armament control panel. Nozzle control lever on the left console just forward of the throttle is in position corresponding to nozzles full aft for wing-borne flight.

Pilots strapped in awaiting a call to launch can simultaneously monitor two radio frequencies since the aircraft is equipped with two separate Collins AN/ARC-182(V) UHF/VHF-AM/FM radios. The VHF-FM capability is particularly important because ground unit radios operate primarily in this frequency band. While alert aircraft are normally under the control of the direct air-support center, under some conditions helicopters and AV-8s may be placed in direct contact with forward air controllers.

Jackson taxied the aircraft to the duty runway using nosewheel steering, essential because the tandem, bicycle main gear arrangement precludes differential braking. The 21,550-lb.-thrust Rolls-Royce F402-RR-406A Pegasus 11 engine has high residual thrust at idle rpm, and the flight manual recommends deflecting the nozzles 45-60 deg. when taxiing to reduce any requirement to ride the brakes. The brakes are designed for the limited requirements of the short takeoff/vertical landing (STO/VL) regime and are thus more prone to overheating than brakes on conventional aircraft.

The twin-spool, axial-flow high bypass turbofan engine with thrust-vectoring nozzles is the heart of the aircraft and was designed specifically for the demands of STO/VL operations. The low-pressure spool is a three-stage overhung fan driven by a two-stage low-pressure turbine; the high-pressure spool is an eight-stage high-pressure compressor driven by a two-stage high-pressure turbine. The spools are coaxial and counterrotate to minimize gyroscopic effects. There are no inlet guide vanes, which eliminates the requirement for an engine anti-icing system. The engine has a water-injection cooling system that enables engine rpm to be increased for a given turbine entry temperature to provide increased thrust for vertical operations at high ambient temperatures.

The engine airflow is split between two pairs of mechanically linked rotatable nozzles that exhaust aft for conventional flight and downward for hovering. Fan bypass air is ducted to the forward nozzles, often referred to as the cold nozzles, and the core exhaust is ducted to the rear nozzles—the hot nozzles. Nozzle position is controlled manually by the pilot using a lever on the throttle quadrant. The nozzles can be rotated from zero deg.—full aft—for conventional flight to 82 deg. for hovering and up to 98.5 deg. for braking with reverse thrust.

The Harrier relies on conventional aerodynamic flight controls during wingborne flight but must use a reaction control system at low speeds and when hovering. The system consists of six puffer valves located at the nose, tail and wing tips that exhaust eighth-stage, compressor-bleed air to provide roll, pitch and yaw control. A butterfly valve automatically routes air to the puffer valves based on nozzle position.

The Pegasus 11 engine in the AV-8A and AV-8B has undergone continuous reliability and maintainability improvements. The basic F402 in the AV-8A was followed by the F402-RR-404 version installed in the AV-8B full-scale engineering development aircraft, which in turn was followed by the F402-RR-404A with increased-capacity bearings and an improved intermediate case for follow-on production aircraft. The F402-RR-406 incorporated improved hot-section cooling, and the current F402-RR-406A is equipped with a Dowty-Smiths digital engine control system that replaces the hydromechanical fuel control used on all the previous engines.

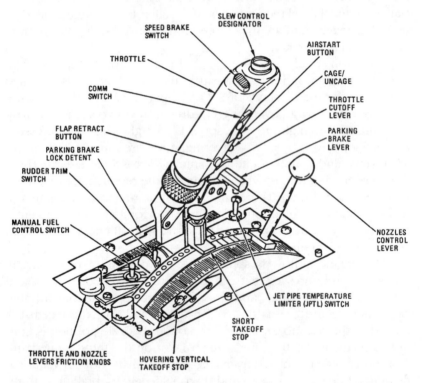

TAV-8B throttle nozzle quadrant has normal communications and speed brake switches on the throttle plus controls for the flaps and the Hughes angle rate bombing system—the slew control designator on the top of the throttle. The nozzle control lever is in the 0-deg., full-forward position. Pulling the lever aft rotates the nozzles down for hovering flight. Various stop positions are available to ease positioning the nozzles to standard settings for short takeoffs and for hovering.

An improved inlet on the AV-8B that incorporates a single row of auxiliary inlet doors to increase airflow at slow speed has improved thrust by 300 lb. In addition, lift improvement devices (LIDS)—two longitudinal strakes and a retractable forward fence on the lower fuselage—capture the reflected jet exhaust during takeoff and landing to further increase thrust by approximately 1,200 lb.

Rolls-Royce is developing an F402-RR-408 version that will extend the time between hot-section inspections from the current 500 hr. to 1,000 hr. and increase the thrust by approximately 3,000 lb. The fan-pressure ratio of the low-pressure compressor will be increased and the fan will be strengthened against bird strikes. New materials and new cooling techniques will be used in the low- and high-pressure turbines.

Takeoff information

Jackson had selected the short takeoff (STO) display on the digital display indicator (DDI) and, after I entered the runway temperature, altimeter setting, field elevation and aircraft gross weight, the DDI showed the following information for a dry takeoff:

- Nozzle rotation airspeed—85 kt.
- Nozzle setting for STO—55 deg.
- Ground roll-400 ft.
- Distance to clear 50-ft. obstacle—550 ft.

Our aircraft weighed 22,800 lb. as we lined up for takeoff—basic aircraft weight of 15,300 lb. plus 7,000 lb. of fuel and 500 lb. of water. Runway temperature was 92F and we set up for a short takeoff. Jackson set the STO stop on the nozzle quadrant to 55-deg. and gave me control of the aircraft. Even with the high runway temperature, the aircraft had almost a 1:1 thrust-to-weight ratio and accelerated rapidly as I advanced the throttle. I used the head-up display to monitor the airspeed, and at 85 kt, I pulled the nozzle control lever back to the stop and we were airborne in about 5 sec, or less. Jackson got the gear up, guided me through the procedure of "nozzling out"—easing the nozzles forward from 55 deg. to 0 deg. as the aircraft accelerated—and I climbed out on a course for the restricted area at the Naval Air Test Center near Patuxent River.

Jackson had dialed in the NAS Patuxent River Tacan and, after I had selected Tacan on the DDI, I put the aircraft on the heading displayed. In addition to the normal mileage readout, the display also included the time to the station at present airspeed. I then switched to the cruise mode, and inertial navigation information was displayed with Patuxent River as the first waypoint. Jackson had used the coordinates of the Tacan station for the waypoint position and the information displayed was virtually identical. For someone used to little more than a compass and an elapsed time clock for low-level dead-reckoning navigation, it was an impressive display.

The TAV-8B is a "head-up" aircraft in all phases of flight, and the HUD symbology allows you to fly it that way.

In the navigation mode, symbology includes heading, airspeed and altitude information superimposed on a flight path/pitch ladder and the aircraft's velocity vector. Angle of attack, Mach number and aircraft g-forces are displayed in digital format on the lower left side of the display, while rate of climb information and the distance to the selected waypoint are presented on the lower right side. A heading marker on the azimuth scale indicates the heading to the selected waypoint.

In the vertical/short takeoff and landing (VSTOL) mode, the HUD displays information more appropriate to that flight regime, including digital jet pipe temperature, engine rpm, sideslip and flap and nozzle position. The displays also shift the angle of attack and rate of climb data to an analog display, indicating trends more clearly.

Relatively little reference is required to the instruments inside the cockpit. Radios and IFF are tuned and DDI entries are made using the coordinate control panel directly below the HUD. The push-button armament control panel just below the DDI is easily accessible.

After getting clearance into the restricted area I started some hard turns left and right to get the feel of the aircraft. The controls were smooth and the aircraft had a good feel under g-loading. The AV-8B has leading-edge root extensions (LERX), developed by British Aerospace, that increase pitch rate and lift and provide improvements in

U.S. Marine Corps McDonnell Douglas-British Aerospace TAV-8B Harrier 2 full-scale engineering development aircraft is shown hovering. The nozzles are deflecting engine thrust downward. Forward starboard nozzle is visible as are the longitudinal strakes and retractable fence that make up the lift improvement device system.

the aircraft's instantaneous turn rate and handling qualities at high angles of attack.

I followed the turns with a series of wingovers and then set up for a loop. We entered at 450 kt. at an altitude of 10,000 ft. I added full power and started a 4g level pull-up using the HUD as a reference for both initial g-force and wing position. As we went over the top at 190 kt., I shifted to the angle of attack readout on the HUD and eased the throttle back to about 85%. At this point I was flying the aircraft more by feel since my scan pattern at this stage was not up to assimilating all the information available. I had not kept a completely wings level attitude throughout the maneuver and, when I added power and came back to level flight, we had drifted a bit off heading but had managed to stay in the restricted area.

Simulated low-level attack
Next we dropped down to 500 ft. over the Chesapeake Bay for a series of maneuvers designed to simulate a low-level run-in and a pop-up attack. Under full power the aircraft quickly accelerated to 550 kt. and then maintained this speed at a power setting of 95%. There was considerable haze on the deck and the visibility was only about 5 mi. with a poorly defined horizon, so we broke off the simulated weapons-delivery portion of the attack after rolling wings-level in a 10-deg. dive.

The TAV-8B is a pure V/STOL trainer at this stage and the Marine Corps wanted it as light as possible in order to improve its vertical performance. At a basic aircraft weight of 15,300 lb., it is already 1,600 lb. heavier than the AV-8B and that weight penalty translates to less fuel available for training. As a result, several systems that are standard in the single-seater have been deleted from the aircraft, although the wiring provisions remain and the systems and equipment could be added if the Marine Corps decides to use the aircraft in a tactical role. Any decision to do this is probably contingent upon acquiring the higher-thrust F402-RR-408 engine. Systems not currently installed in the TAV-8B are the Hughes AN/ASB-19 angle rate bombing system, General Electric's 25-mm. gun pod, a refueling probe and electronic countermeasures.

In addition, the TAV-8B has only two store stations, as opposed to six on the AV-8B, and will normally carry only practice bombs and rockets or two external fuel tanks during training.

Pilots undergoing training at VMAT-203 are making extensive use of the AV-8B Weapons Tactics Trainer to attain proficiency in the aircraft's attack, navigation and ECM systems. The trainer uses an Evans and Sutherland CT-5A visual system, and McDonnell Douglas pro-

vides the flight dynamics and handling qualities software. The AV-8B
weapons systems are controlled by switches on the stick and throttle,
while information is presented to the pilot on the HUD. Pilots emphasized that the only way to develop the manual dexterity and reaction time required is by executing repetitive attacks in the trainer
(AW&ST Sept. 16, 1985, p. 54).

Restricted site operations
We pulled up from our simulated attack and headed back for Quantico using the inertial navigation system to point the way. Jackson
took the aircraft as we approached the Davis Crossroads area and set
up for a series of approaches to a confined area to demonstrate the
aircraft's capabilities to land and operate out of restricted sites. He
first simulated a conventional aircraft approach at 128 kt. and 30-deg.
nozzles to set the stage as a reference. On the second approach he
went to 65-deg. nozzles and lowered the approach speed to 88 kt.—
what Harrier pilots call a slow landing. On the final approach, he de-
celerated to a full hover and then accelerated for our return to
Quantico.

In the hover, I followed along on the controls while Jackson flew
the aircraft. Several factors have combined to make the AV-8B much
easier to fly than the AV-8A. The AV-8B's larger, supercritical wing
with automatic maneuvering flaps-and drooped ailerons in the vertical regime—is still generating lift at 42 kt., while the AV-8A wing loses
lift at 96 kt. The longer moment arms on the AV-8B's reaction control
system provide more authority. Most importantly, a stability augmentation and attitude hold system (SAAHS) that is operational throughout the entire envelope has been installed.

We came back into the break at the Quantico airfield and made a
final vertical landing with 1,200 lb. of fuel. The outside air temperature
was up to 94 deg., and Jackson turned on the water to provide a safe
margin for the engine, although he could have made a dry landing
without exceeding engine temperature limits. The single-seat AV-8B
can carry up to 9,200 lb. of ordnance but would have to make a short,
rolling takeoff to lift this load. With a typical close-air support load of
4,000 lb. and full internal fuel of 7,700 lb., an AV-8B at takeoff would
weigh 25,400 lb., considerably greater than its maximum vertical takeoff weight of approximately 19,000 lb. But the aircraft could get airborne with this load in less than 1,000 ft. and could then return and
land vertically, which is how the Marine Corps intends to employ it.

I reflected on the Harrier's capabilities compared with the Mc-
Donnell Douglas A-4Es flown by Marines out of Chu Lai, Vietnam, in
1965. Initial operations from a 4,000-ft. aluminum-matting runway—

many tons of which had been transported by ship—with about 3,500 lb. of ordnance required that the A-4s launch with reduced fuel loads using jet-assisted takeoff (JATO). The A-4s then had to top off from refuelers overhead the airstrip after circling over the South China Sea to jettison the JATO bottles. Two sets of heavy arresting gear had to be installed to handle aborts and for the arrested landings that were required on every recovery. A liquid oxygen plant was required. To conduct sustained operations and eliminate the requirement for JATO launches and arrested landings, the strip was eventually extended to 8,000 ft., and another set of arresting gear was installed at midfield.

If AV-8Bs had been available, a few thousand feet of matting would have been enough, and a road might have been found that would have precluded transporting even this modest amount of matting. The Marines might even have operated from the short grass strip on the nearby small island of Cu Lao Re and remained less vulnerable to ground attack.

Mirage 2000 fighter combines acceleration, low-speed stability

Robert R. Ropelewski/Istres, France
June 24, 1985

Avions Marcel Dassault-Breguet Aviation Mirage 2000 fighter, now in its early operational stages with the French and Indian air forces, combines exceptional acceleration, agility and low-speed stability characteristics with well-planned multifunction cockpit displays to produce a versatile combat aircraft that compares favorably with other new Western fighters in this category.

Like its primary competitors, the General Dynamics F-16, the McDonnell Douglas F/A-18 and the Northrop F-20, the Mirage 2000 incorporates the best of recent aerodynamics, electronics and propulsion technology to achieve impressive gains in performance, maneuverability and overall handling qualities.

Familiarization flight

I found this to be the case when I flew the Mirage 2000 here recently on a 1-hr. familiarization flight. In addition to a nicely tailored flight control system, the aircraft is equipped with an informative, low-workload integrated cockpit display system and head-up display that allowed it to be flown purposefully and precisely even on a first flight in marginal weather conditions.

Istres is the primary French flight test center, and most of Dassault-Breguet's flight test activities, are centered here. I flew the No. 1 two-seat Mirage 2000B with Dassault project pilot Guy Mitaux-Maurouard, who has been flying the Mirage 2000 since the first flight of the first prototype in March, 1978 (AW&ST Mar. 20, 1978, p. 16). The aircraft was powered by the new Snecma M53-P2 turbofan engine rated at 21,340 lb. (9,700 kg.) sea level takeoff thrust in afterburner and 14,432 lb. (6,560 kg.) thrust in military power (maximum power without afterburner). In comparison, the M53-5 engines powering initial production versions of the Mirage 2000 are rated at 19,840 lb. in afterburner and 11,100 lb. dry.

Preflight preparations consisted of a briefing on the aircraft and its characteristics and a 1-1.5 hr. session on a desktop trainer, which provided some familiarization with the Mirage 2000's various controls and displays. The trainer was originally developed for in-house use but is now being adopted by the French air force to provide continued training for squadron pilots and to prepare them for specific missions.

I occupied the rear cockpit for the flight, and except for starting the engine and for some brief demonstrations conducted from the front seat by Mitaux-Maurouard later in the flight, I was able to operate the aircraft entirely from the back seat.

Although the rear seat of the Mirage 2000B is raised slightly compared with the forward seat, it is still not high enough for the pilot in the rear cockpit to have a clear field of vision through the rear headup display. As a result, the rear HUD is simply a repeater of the front cockpit HUD. In reality, I quickly forgot this because the effect is that of looking directly out the front of the aircraft.

Forward visibility

Using the HUD repeater as I taxied to Runway 34 for takeoff, I had the impression that I was looking directly at the taxiway immediately in front of the aircraft. The same impression of excellent forward visibility lasted throughout the flight. Sideward and rearward visibility also were very good, impaired only slightly by narrow canopy bows just in front of each pilot.

Mitaux-Maurouard had used external electrical power to start the M53 engine on the flight line, primarily because of the extensive amount of flight test-equipment still installed in the No. 1 Mirage 2000B that needed to be operating prior to engine start. Normally, the aircraft can be started using its own internal battery power. The engine start was accomplished quickly and without difficulty, and we were cleared to taxi shortly after Mitaux-Maurouard finished a series of checks of on-board flight test and telemetry equipment. Aircraft weight at that point was approximately 11,000 kg. (24,200 lb.).

There is little difference in the weight and performance of the two-seat Mirage 2000 compared with the single-seat version, according to program officials. Incorporation of the second cockpit meant eliminating the twin 30-mm. cannons carried internally on the single-seat aircraft, and fuel capacity had to be reduced slightly from 4,100 liters (1,083 gal.) to 3,850 liters (1,017 gal.). Apart from this, the two-seat aircraft is said to have the same mission capabilities as the single-seat version.

After a rather lengthy taxi to the runway and an extended hold for other traffic, we were cleared onto the runway for takeoff. Once in position, I ran the throttle up to military power for one last check of the engine before releasing the brakes and moving the throttle past the military power detente and into the maximum power/afterburner zone.

The engine spooled up quickly to military power, and afterburner ignition was immediate when selected with the throttle.

Among the useful features of the Mirage 2000 head-up display is a small box in the lower right corner of the display that provides a numerical readout of aircraft acceleration and deceleration on the ground. Mitaux-Maurouard calculated that this box should indicate 0.70-0.75g for takeoff at the aircraft's current weight. In this case, the readout stabilized after a few seconds at 0.75g and stayed there until liftoff. The benefit of this feature is that it eliminated the need to perform time/distance checks using runway length-remaining markers to confirm adequate acceleration.

Although we were carrying no external stores, the full fuel load and the few thousand pounds of flight test equipment adequately simulated the weight of a Mirage 2000 on a typical air intercept mission, according to Mitaux-Maurouard. Twelve seconds after brake release the aircraft reached our predetermined rotation speed of 120 kt., and I pulled back gently on the stick to raise the nose to a 13-deg. attitude as commanded by a small aircraft symbol on the head-up display. The attitude commanded by the symbol is determined by the aircraft's mission computer and varies with takeoff weight. The aircraft lifted off a few seconds later at a speed of about 150 kt.

Climbing left turn

I raised the landing gear and kept the nose from rising more than a few degrees until we reached 500 kt. indicated. This occurred about 30 sec. after liftoff. At that point, I pulled the nose up to about 45-50 deg. of pitch and began a climbing left 180-deg. turn toward the Mediterranean coast a few miles to the south.

It was an extremely hazy day, and most ground references were obscured after we had climbed a few hundred feet. All of the necessary information to fly the aircraft safely was displayed simply and understandably on the head-up display, however, and I continued the maximum-rate climb without having to make a transition to head-down instruments.

Still in afterburner, and with a speed of Mach 0.92, I rolled the aircraft inverted to stop our climb at 32,000 ft., which we reached 1 min. 45 sec. after brake release. One minute later we were at Mach 1.5, and I brought the throttle out of afterburner to military power. Because we were still in the clouds at that altitude, I pulled the nose up again to continue climbing until we were in the clear at 42,000 ft. French air force Mirage 2000s used in the air defense role will eventually be equipped with a data link from groundbased air defense radars that will provide steering cues to the crews to intercept intruding aircraft without the need for air-ground voice communications.

Operational ceiling

Primary mission for the Mirage 2000 within the French air force is as an air defense interceptor, and the aircraft has an operational ceiling of approximately 60,000 ft. (18,000 meters) for that reason. It also has a maximum speed in excess of Mach 2.2, but I had agreed with Mitaux-Maurouard that it was not worth the fuel required to reach that speed on this flight.

Our speed at this point was still Mach 1.2, even though I had reduced the thrust to military power at 32,000 ft. and had climbed an additional 10,000 ft. thereafter. Mitaux-Maurouard said the aircraft in the clean configuration was capable of maintaining a speed in excess of Mach 1 without the use of the afterburner. I performed several 360-deg. aileron rolls right and left, and full lateral stick deflections at that speed and altitude produced roll rates of about 180 deg. per sec. Maximum roll rate for the Mirage 2000 is about 280 deg. per second, which I observed later at lower altitudes. The maximum rate in any given circumstance is determined and limited by the flight control computer.

The quadruple-redundant fly-by-wire flight control system used in the Mirage 2000 gives the aircraft a solid yet responsive feel throughout the flight envelope. In this case, yaw limiters in the system kept the nose riveted to a point on the horizon as I rolled the aircraft, indicating an excellent target tracking capability in an air combat situation.

Throttle movements

The Snecma M53 engine is relatively robust, and it was possible to move the throttle from stop to stop without restriction at our 42,000-ft. altitude. Mitaux-Maurouard said the only limit on throttle movements with the M53 engine is at airspeeds below 100 kt. at altitudes above 40,000 ft., where no throttle movements are permitted.

The Mirage 2000 has an impressive turn capability throughout its flight envelope, and this was demonstrated in a series of hard turns I performed with the aircraft beginning at 40,000 ft. At a speed of about Mach 1.05, I rolled into a hard left turn, pulling the stick nearly full aft. The g indicator on the instrument panel peaked at about 4.5g, and the aircraft's angle of attack increased to about 25 deg. The fly-by-wire system uses inputs from the aircraft's air data computer and other sources to limit load factor and angle of attack throughout the Mirage 2000 flight envelope.

The aircraft is normally limited to 9g in the air-to-air configuration, but up to 11g can be attained in an emergency by giving an extra 30 kg. (66 lb.) tug on the stick to overcome an artificial detente in the flight control system. When the aircraft's weapon delivery fire control system is switched to the air-to-ground mode with a switch on the stick, the load factor limits are automatically reduced to 5.5g, with an emergency limit of 7.5g achievable with the additional force on the stick.

Angle of attack limits range from 22-28 deg. on the Mirage 2000 and are imposed automatically by the fly-by-wire flight control system based on flight conditions at the time. In the initial high-altitude turn, the aircraft turned through 90 deg. in about 10 sec. and through 120 deg. in about 14 sec. before I rolled to the right to reverse the turn.

Except for a very slight buffeting and a dull, relatively high-pitched rumble that could be heard from the airflow over the Mirage's delta wing, the aircraft remained very stable, providing good pointing and tracking accuracy. Automatic leading edge flaps are used on the Mirage 2000 to vary the camber of the wing and sustain the aircraft's maneuverability, but the extension and retraction of these flaps were not noticeable in the cockpit.

Stick forces remained uniformly comfortable throughout the flight envelope, and there was no need for any trim changes as the

Single-seat Mirage 2000 in typical French air force air defense configuration carries Matra Super 530 medium-range air-to-air missiles on inboard wing pylons and Matra 550 Magic short-range air-to-air missiles on outboard wing pylons.

Cockpit of the Mirage 2000 is relatively spacious and well organized, with most essential flight, navigation and weapon delivery information presented on the head-up display (top, center) or on the radar screen (center).

aircraft accelerated from subsonic to supersonic and back to subsonic. I continued a series of hard instantaneous and sustained turns while allowing the aircraft to descend to a lower altitude. In the low to mid-30,000-ft. altitude range, the Mirage exhibited an instantaneous turn rate in the 10-12-deg. per sec. range up to about 5g load factor. The aircraft is capable of maintaining steady-state 3g turns at Mach 2 at 36,000 ft., although we did not examine this on the flight.

Having bled off a considerable amount of airspeed during the turns at high altitude, I pushed the throttle into the afterburner range and the aircraft accelerated from 180 kt./Mach 0.6 to 275 kt./Mach 0.82 in 35 sec. at 38,000 ft.

During a descent through 22,000 ft. at Mach 0.9, hard turns right and left produced instantaneous turn rates approaching 20 deg. per second at load factors just above 8g and with an angle of attack of 27 deg.

Maximum instantaneous turn rate at these altitudes is 24 deg. per second. Actual degrees rather than units of angle of attack are indicated in the Mirage 2000 cockpit.

I leveled the aircraft briefly at 14,000 ft. and selected afterburner once again to accelerate from 515 kt./Mach 0.97 to 645 kt./Mach 1.21 in 20 sec. After reducing power to decelerate, I engaged the aircraft's Sfena autopilot using the automatic flight control system control box on the left side of the glare shield. I had selected only the attitude hold function, and thus was able to use the stick to establish the aircraft in a suitable attitude for a descending turn to 12,000 ft. before releasing the stick and allowing the autopilot to hold that attitude.

Mitaux-Maurouard armed the altitude hold mode and selected 12,000 ft. on his own AFCS panel, and the aircraft leveled off smoothly when we reached that altitude on our descent.

Heading selection

The autopilot can be engaged in any aircraft pitch attitude except inverted and at roll angles up to 65 deg. With the autopilot engaged, new headings or pitch attitudes can be selected by pressing the thumb trim switch on the top of the stick in the appropriate direction to move an index in the head-up display to the desired heading and pitch attitude. The autopilot can be disengaged temporarily by a paddle switch on the stick or permanently disengaged by another switch.

Mitaux-Maurouard took the controls briefly to demonstrate the zero airspeed maneuver he has been performing in the Mirage 2000 at the Paris and Farnborough air shows in recent years. Starting from an altitude of 8,000 ft. and an airspeed of about 410 kt., he raised the nose to a pitch attitude of about 75 deg. while reducing the throttle to idle. The aircraft climbed to about 13,000 ft. and the airspeed slowed to essentially zero before the nose pitched gently downward to a 90-deg. nose-down attitude and airspeed began to increase. Elevons were almost immediately effective, and the aircraft could be rolled about its longitudinal axis even before the airspeed readout appeared on the head-up display as our speed climbed above 65 kt.

Forward stick pressure

With the aircraft stabilized at 8,000 ft. and 300 kt., I took the controls once again and pulled the nose up 75-80 deg. to execute a similar maneuver. It was necessary to apply just a small amount of forward pressure on the stick to keep the aircraft from going beyond 90 deg. nose-up pitch and ending up on its back. We reached zero airspeed at about 11,000 ft., and again the nose pitched gently downward, airspeed began to climb, and control effectiveness could be demonstrated even at very low speed by rolling the aircraft around its longitudinal axis.

In this and other low-speed situations throughout the flight, there was never an impression of sloppiness in the controls as is often the case at slow speeds in most aircraft. Control feel and response always seemed to be crisp and positive, no matter what the speed. Additionally, stick forces required throughout the flight seemed to be especially light, probably because the angle of attack and g-limiters in the flight control system eliminate the need for heavy stick forces to prevent overstressing the aircraft.

All of the switches necessary to operate the radar, head-up display and weapons systems on the Mirage 2000 in combat conditions are located on the stick and throttle so that no distracting hand or eye movements are necessary in critical circumstances. Mitaux-Maurouard switched the HUD and radar displays from the navigation mode to the air-to-air mode to take a brief look at the radar in that application.

Current production Mirage 2000s are equipped with a Thomson-CSF RDM multifunction X-band monopulse Doppler radar. Basic control switches for the radar are on a panel on the left console of the Mirage cockpit, just to the left of the throttle. In high-intensity combat situations, all of the appropriate modes for the radar can be selected with a rocker switch on the front of the throttle grip. The RDM radar has a target detection/acquisition range of about 50 naut. mi. against targets with a radar cross section of 5 sq. meters (53.8 sq. ft.) or more flying above the horizon. But because the radar is a compromise between both air-to-air and air-to-ground requirements, its detection range is reduced to about 20 naut. mi. for low-altitude targets hidden by ground clutter. There were no distant fighter-size targets to acquire during our flight, although the radar easily acquired and tracked a commercial aircraft flying at 30,000 ft. at a distance of 35 naut. mi.

The radar antenna could be easily controlled using the switches on the throttle to search for and designate the target. Head-down radar display on the center instrument panel directly below the HUD facilitated use of the radar without excessive eye or head movement. The radar display showed the altitude and heading of the target aircraft as well as its range and relative bearing from our position. The RDM radar has a track-while-scan capability, and Mitaux-Maurouard selected that mode to allow the system to look for other targets while tracking the transport aircraft.

RDI radar

A more specialized air defense radar for French air force Mirage 2000s is under development by Thomson-CSF and should become operational with French air force squadrons in early 1988. The new radar, designated the RDI, is a high pulse repetition frequency (PRF) pulsed Doppler system intended specifically for air defense missions against fast-moving targets. The radar is expected to have a significantly improved look-down target acquisition capability compared with the current RDM radar.

We terminated the air-to-air radar tracking, and I turned the Mirage back toward the French coastline and Istres. I descended to 1,500 ft. while Mitaux-Maurouard switched the radar to the air-to-ground mode. The RDM radar has several air-to-ground functions, including:

- Ground mapping with Doppler beam sharpening.
- Image freeze.
- Contour mapping for terrain avoidance and blind penetration.
- Air-to-ground ranging.
- Ground moving target indicator (GMTI).

Terrain delineated

The contour mapping mode with a 60-mi. range was ideal for the limited visibility in which we were flying and clearly showed areas of rising terrain in the hilly region around Istres. Terrain that was at or above our altitude of 1,500 ft. was shown in red on the radar display, while all else was depicted in green, showing the safest path to fly to avoid collision with the ground.

Mitaux-Maurouard selected the air-to-ground ranging mode on the radar, and we executed several simulated attacks against targets of opportunity in the French countryside. The armament control panel at the lower left corner of the front instrument panel could be reconfigured from one type of weapon to the next with the push of a button, and we were able to transition quickly from a simulated rocket firing attack to a high-drag bomb drop to a clean bomb delivery to a cannon pass.

The head-up display provided clear, unambiguous steering cues and, in the case of the bombs, offered the option of using either continuously computed impact point (CCIP) or continuously computed release point (CCRP) delivery modes.

Low-level ride quality of the Mirage 2000 was extremely smooth, even over hilly terrain on a relatively hot day. This further enhanced the excellent pointing accuracy of the Mirage, leaving the impression that it would compete very favorably with most dedicated ground-attack aircraft in terms of air-to-ground weapons delivery accuracy.

Our final attack completed, we climbed to 4,000 ft., where Mi-taux-Maurouard demonstrated the low-altitude maneuverability of the Mirage 2000 by rolling the aircraft inverted and pulling the stick back to execute a split-S maneuver. The airspeed was 220 kt. at the start of the maneuver, which was completed at an altitude of 1,700 ft.

I took the controls and executed a similar maneuver with the same results. Only about 5g was needed to accomplish the maneuver in that limited airspace, and it was evident that an even tighter pattern was possible.

With the Thomson radar in the ground mapping mode and with Doppler beam sharpening, the runways at Istres—then about 10-12 mi. away—were clearly distinguishable on the radar display. Mitaux-Maurouard moved the target designator on the radar display to the initial point for entry to the Istres traffic pattern, and I turned the aircraft to fly to that point at an altitude of 2,000 ft.

The Mirage 2000 autopilot is equipped with an automatic landing mode that can be coupled to a ground-based instrument landing system, and we had planned to execute an automatic ILS capture and approach on our return to Istres. However, the ILS transmitter at the base was shut down for maintenance when we returned, and we were unable to observe the autopilot in the automatic landing mode.

Back at the field, I performed a touch-and-go landing before returning for a fullstop landing to end the flight. Stability of the aircraft together with the useful information provided by the head-up display made the aircraft easy to land, even though it was my first flight in it, and I had not received any previous simulator preparation.

Angle-of-attack range

Both approaches were made using only the HUD for a reference. In addition to the velocity vector symbol, which showed the flight path of the aircraft, the HUD also displayed two brackets—one on each side of the display—which defined the acceptable angle-of-attack range (13-15 deg.) for the approach. These brackets appeared when the Approach mode was selected on the armament control panel. Keeping the velocity vector between the upper and lower limits of these brackets and over the desired touchdown point on the runway at the same time resulted in an optimum approach to landing.

Although I was not at all cognizant of airspeed during these approaches, our final approach speed was about 140 kt. and touch-down speed was about 125 kt. Dassault test pilots said the Mirage 2000 can be flown safely and comfortably down to a speed of 100 kt. Mitaux-Maurouard earlier had suggested that I execute an Immelmann maneuver after the first touch and go, igniting the afterburner and pulling the nose straight up and over to end up with the aircraft on its back at 1,500 ft. before rolling right side up with the aircraft pointed downwind. However, our fuel state was such that the use of the afterburner was not practical and I entered a conventional downwind pattern for a final landing.

I used maximum braking after touchdown on our final landing. The Mirage 2000 has a heavy-duty antiskid brake system, and I was able to bring the aircraft to a full stop in what appeared to be less than 2,500 ft. from the touchdown point. Distance markers beside the runway showed nearly 3,000 meters remaining on a 3,800-meter runway.

The aircraft was shut down in the chocks with approximately 380 liters (100 gallons of fuel remaining). The Sagem navigation/weapon delivery system showed a 0.5-naut.-mi. error after a flight of 55 min.

Dassault has received firm orders for 275 Mirage 2000s so far, including 113 for the French air force, 20 for Egypt, 40 for India, 26 for Peru, 36 for Abu Dhabi and 40 for Greece. The French air force now has two squadrons operating, and the first aircraft were delivered to India last September. The company is rolling out five to six Mirage 2000s a month but is building toward an eventual rate of 10 aircraft a month.

Flight testing is still under way on the air defense version equipped with the air-to-air optimized RDI radar and Matra Super 530D medium-range air-intercept missiles. Program officials estimated that this flight development work was about two-thirds completed. The Super 530D has been tested against drone targets up to Mach 2 and 60,000 ft., but remains to be tested against a Mach 3 target at 80,000 ft. and against high subsonic targets at altitudes down to 100 ft.

Once the higher-thrust M53-P2 becomes available, all export Mirage 2000s are expected to be equipped with that engine. The same engine also will equip the Mirage 2000N, the low-altitude penetration/interdiction version of the fighter that is now in development and flight test.

The RDI radar, specifically optimized for the French air force air defense mission, is expected to enter service on Mirage 2000s delivered in about two years. For the time being, that radar will not be available to export customers.

In any case, Dassault officials believe most export customers for the Mirage 2000 have a need for a multimode radar such as the current RDM rather than a specialized system like the RDI.

The Mirage 2000 is being produced with an integral, internally mounted electronic countermeasures system but has been qualified to carry a variety of externally mounted ECM pods and externally carried electronic intelligence (elint) pods as well.

Navy F/A-18 demonstrates dual-mission performance

David M. North/St. Louis
August 11, 1980

U.S. Navy's McDonnell Douglas F/A-18, evaluated by this *Aviation Week & Space Technology* pilot during a 1.3-hr. flight here, offers the service a high-performance aircraft platform with an advanced avionics package capable of performing the fighter and attack role with one pilot and is still an easy aircraft to fly.

Although the F/A-18 has been under attack on technical and budgetary grounds, with even some consideration of program cancellation (AW&ST June 30, p. 18), no operational or performance problems were evident from the F/A-18's cockpit when I flew the Navy's and Marine Corps' newest fighter/attack aircraft late last month.

What was evident from the preliminary briefings and flight from the McDonnell Douglas facility here at Lambert Field that the F/A-

Two U.S. Navy F/A-18s prepare for a catapult launch from the deck of the USS Constellation.

18 had performance comparable to other front line fighters, had agile engine performance and holds the potential for increased aircraft reliability and maintainability the service needs in the 1980s and beyond. This last point becomes more important as the Navy faces shortages of trained personnel to maintain the fleet aircraft in its inventory.

The F/A-18 offers the Navy other attributes for the future. It is a single-pilot aircraft, at a time when the service's retention rate of crewmembers is low, and allows the option of transferring attack pilots to fighter billets, and in the opposite direction, while flying the same aircraft. The fact that the F/A-18 is proving to be a forgiving aircraft and relatively easy to fly also should make the aircraft safer for pilots who do not get to fly the aircraft on a regular basis.

Both Rep. Jim Lloyd (D.-Calif.), who flew a two-seat F/A-18 from the Navy's Patuxent River, Md., test facility late last month (AW&ST Aug. 4, p. 15), and I agree that one of the hardest tasks for novice F-18 pilots is knowing what button to push for the numerous displays available on the aircraft's three cathode ray tubes.

Prior to the flight with McDonnell Douglas director of flight operations John E. Krings, I had flown the F/A-18 simulator in late 1976 for a total of 1 hr. to evaluate the ability of one pilot to fly the aircraft and operate all the aircraft's weapons systems (AW&ST Jan. 31, 1977, p. 64). On the morning before the flight, another 45 min. was logged in the flight hardware integration simulator to reacquaint myself with the F/A-18's cockpit.

The cockpit displays have not changed significantly over the past three years. Two cathode ray tubes are located on either side of the communication, navigation and identification panel. The head-up display is mounted above the CNI, or up-front control, and the moving map display is to be mounted below it. Until McDonnell Douglas is able to acquire enough horizontal situation displays to equip its aircraft, it has put another cathode ray tube in its place.

The standby instruments, such as airspeed, rate of climb, altimeter and a small attitude gyro are on the lower right portion of the instrument panel. The digital engine instruments containing exhaust gas temperature and turbine speed are on the left lower instrument panel, as is the nozzle position indicator.

One change incorporated in the simulator and the aircraft is the use of dedicated warning lights for certain system malfunctions such as hydraulic power that would need to be known prior to engine start. This warning light panel is located below the hook handle on the lower right side of the instrument panel.

The aircraft flown was TF-2, the second two-seater version of the 11 F/A-18s built for the research and development phase of the Hornet program. The other nine aircraft are single-seat aircraft. Prior to the flight here, TF-2 had accumulated 198 flights for approximately 230 hr. of flight time. The 200th flight for TF-2 occurred on the same afternoon I flew the aircraft.

Following a flight briefing by Krings, and a thorough checkout of the Martin Baker ejection seat by Robert Lindsay, supervisor of the pilot's equipment laboratory, I strapped myself into the TF/A-18's rear seat. The instrument panel of the rear seat is almost identical to that of the front seat. Missing from the rear seat are the gear handle, flap control, engine shutoff switches and a head-up display. Throughout most of the flight the head-up display presentation was shown on the left cathode ray tube in the rear cockpit.

Krings started the aircraft's auxiliary power unit with battery power. Once the APU ready light indicated sufficient power was available for engine start, the right engine and then the left engine were started. Exhaust gas temperature peaked at about 600C and settled down to 410C at idle power. The APU automatically shut down once both engines were at idle power and the unit had time to cool. Throttles had been brought to idle at the first indication of engine rpm.

Displays in both cockpits are the same, and either pilot can choose the display to be shown on the cathode ray tubes at any time during the flight. Krings initially checked the engine performance display, then the mission display presentation that showed the status of most systems on the aircraft and of some that had not been installed yet, and finally the flight control display.

The flight control display gives an indication of the status of all four channels of the fly-by-wire system. Trim, stabilizer and flap status also are indicated on the flight control display.

During the pretaxi check list, only one maintenance item could be found: although the aircraft had 9,000 lb. of fuel the flight display CRT and the audio warning intermittently indicated that the fuel on the aircraft was at the "bingo" level, or the point at which the aircraft should be headed for its home base during a flight. The fuel gauges read 9,000 lb. of fuel.

The low residual thrust of the two General Electric F404-GE-400 turbofan engines necessitated adding power to taxi the aircraft on the 90F day. This same low thrust at idle power became an asset when slowing the aircraft in flight and after landing.

Prior to takeoff, the flight display cathode ray tube indicated the nose wheel steering was in the taxi and takeoff low response mode and the canopy was closed. The takeoff was made from an intersection of Lambert's Runway 30 to avoid waiting in line for numerous commercial transport aircraft. With a full afterburner takeoff made by Krings, the F/A-18 was airborne in 1,200 ft. at a speed of approximately 125 kt. The aircraft's nose initially was rotated to a 30-deg. pitch angle until 250 kt. was attained and then rotated to 37 deg. on the CRT's attitude display.

The top of the 8,500-ft. terminal control area was reached approximately 1 mi. from the close end of the runway. Krings said it is "an enviable way to leave an airport."

The aircraft was turned to the south of St. Louis to work in one of two areas McDonnell Douglas uses for its test flying. I took command of the aircraft at slightly above 10,000 ft., while indicating 300 kt. and a 6,300-lb./hr. total fuel flow.

The visibility from the back seat was good during 30-45-deg. banked turns that were made to get a feel for the aircraft. The TF/A-18 was responsive to both pitch and roll inputs. Force required on the standard-mounted center stick was higher than found in other aircraft I have flown, and according to Krings, higher than the F-15. This gave

The McDonnell Douglas F/A-18 has become the mainstay of the U.S. Navy carrier operations and has been purchased by other countries, including Canada, Australia, Spain and Kuwait.

the impression that the F-18 has more solid handling characteristics. The aircraft also gave the impression of being neutrally stable. Once it was trimmed and the nose was placed at some point on the horizon, it tended to remain at that point.

TF-2 was restricted to 20 deg. angle of attack, 4-5gs and 535 kt., primarily because the aircraft was scheduled to enter update modifications in the next few days. Administrative procedures to clear the aircraft into a larger part of the envelope had not been accomplished. TF-2 retained the earlier wing leading edge extension configuration (AW&ST June 2, p. 22). Other F/A-18s in the test program have been cleared out to the edges of the performance envelope.

To date in the test program, the F/A-18 has achieved 692 kt. and Mach 1.9, while the current Navy requirements call for 650-kt. maximum and Mach 1.7 maximum. These figures are for Naval Preliminary Evaluation and can be revised to meet changing requirements.

Preliminary angle-of-attack requirement for the F/A-18 is from –6 deg. to +35 deg. According to McDonnell Douglas officials, the aircraft has achieved a –8-deg. to +82-deg. angle of attack. During one test flight the aircraft was flown at an 80-deg. angle of attack, with one engine in afterburner and the other in idle and did not depart from controlled flight.

Flight test pilots have found that the F/A-18 has no tendency to depart from controlled flight in almost any segment of the flight envelope. When a sideslip has been induced, the aircraft has been found to self-correct. Recently accomplished in this area is the use of maximum cross deflection of the controls for 15 sec. to put the aircraft into an aggravated angle of attack and sideslip condition with no resultant departures and spins.

Another 20 flights are still required in the test flight program to finish the extremes of g force flight. The Navy requirement is for a –2 to a +7g regime. The aircraft already has demonstrated a –2.8 to a 7.7g performance, but the goal is for a –3 to a +8g standard.

At 12,000 ft. and approximately 300 kt., the engine throttles were advanced to maximum military power and speed was allowed to build up to 400 kt. for a climb. Within 45 sec. the aircraft was passing through 20,000 ft. In another 2 min. the aircraft was at 31,000 ft. and the throttles were pushed straight forward for afterburner power. At 39,000 ft. the aircraft was indicating Mach 0.93. It had taken a total of 3.2 min. to reach this speed and altitude. The aircraft went supersonic in 3.5 min. and reached Mach 1.3 in 4.5 min. from beginning the climb.

Two characteristics of the aircraft were quite noticeable during the climb and later flight. The first was that the cockpit was very quiet at Mach 1.3 and at high speeds at lower altitudes. A hot mike was used during the entire flight with no noise discomfort. In conjunction with this quietness, there were virtually no audio clues that the aircraft was traveling either slow or fast, accelerating or decelerating, or that g forces were being pulled.

The digital flight control system compensated for pitch so smoothly it went unnoticed unless the flight control display was monitored on the CRT to discern pitch changes. This also was less evident in the roll mode. Later, when the speed brakes, landing gear and flaps were extended, there was no aircraft pitch change noticeable. Trim button on top of the control stick was used sparingly throughout the flight.

The power was eased back to military power, and as the aircraft was decelerating, 2g turns were accomplished and then 3g turns. There was no buffeting as the relatively tight turns were made and the speed fell below supersonic.

According to one McDonnell Douglas test pilot who has flown both the F-15 and the F-18, the F-18 has the advantage in the dogfight regime initially because of its fast engine spool-up time, while the F-15—with its higher thrust-to-weight ratio—has a longer term edge. The test pilot said in air combat it would be competitive with the F-15 and the General Dynamics F-16.

McDonnell Douglas officials said once all the fixes are incorporated in the wing, acceleration from Mach 0.8 to Mach 1.6 will be cut by 3-4 sec. and that the acceleration from Mach 0.8 to Mach 1.2 will be cut a fraction, but the final figures will not be what they had hoped for. McDonnell Douglas officials point out, however, that in drag races from a slow speed the F-18 will out-accelerate the F-4, the F-15 and the Grumman Aerospace F-14 until higher speeds are reached, when the other aircraft catch up with the F/A-18. Rep. Lloyd confirmed this during his flight in a race with an F-14. After 2 min. from an even start his F-18 was 80 kt. faster although it later was overtaken.

During the *Aviation Week & Space Technology* flight, Krings used excess speed to pull the aircraft up to 45,000 ft. When the aircraft was at an 8-deg. angle of attack and at 165 kt., Krings took the right throttle and advanced it from idle to maximum military power and then back to idle power and back to maximum again without a noise or protest from the engine. By the time this maneuver was completed the aircraft was 47,000 ft. and indicating 116 kt.

Both Navy and McDonnell Douglas officials give high marks to the F404 engines powering the F/A-18 for both performance and reliability. More than one McDonnell Douglas official has admitted that the engine removal rate for the F-15 during its development was considerably higher than the low rate the F-18 is experiencing, and that if the Air Force program had been given the scrutiny the F-18 is now receiving, there would have been more outside pressure on the program.

"It sure makes a test program, such as F/A-18, much easier when you do not have to worry about engine problems," a McDonnell Douglas test pilot said.

Following the throttle response demonstration, both F/A-18 engines were put at 94% engine speed for military power and a descent to 24,000 ft. was initiated. During the descent, various g forces were put on the aircraft and varying angles of attack were tried. Mild buffet in the airframe was encountered at 11 deg. angle of attack, with the buffet increasing slightly as the angle of attack increased beyond 12.2 deg. At an airspeed of 220 kt. the turning radius of the F/A-18 was very tight. Because of the engine thrust and maneuverability of this class of fighter, acrobatics that would not be tried in earlier aircraft until a speed of 400-450 kt. was reached can be accomplished in the 250-kt. regime.

Twice during the descent as the airspeed had built up, roll rate performance of the aircraft was evaluated. Although the roll rate was not timed, the forces exerted by the roll were enough to snap my head to one side. The first F/A-18 to incorporate the modifications attempting to correct the less-than-desired roll rate of the aircraft is scheduled to fly in September. In earlier tests the roll rate was found to be 100 deg./sec. at Mach 0.9 at 10,000 ft., which is 80 deg./sec. less than specified.

At 24,000 ft. and 220 kt., Krings retarded the throttle to shut down the right engine. At 45% engine rpm, the throttle was brought back to idle and an immediate engine relight was obtained. Slowing the aircraft, the right engine was shut down again, and this time the engine speed was allowed to dwindle to 17%. To demonstrate the difficulty of air starting an engine at low rpm, Krings brought the throttle back to idle, which resulted in a low rate of increasing engine rpm and exhaust gas temperature that rapidly increased to 550C and would have resulted in a hot start if the throttle had not again been brought to the cutoff position.

As Krings said, there are many aircraft that at this point would have to dive off altitude to increase engine rpm or would have to return to base with one engine. He then used the engine start switch, which ducted engine air from the left engine to the right engine and a normal engine start was obtained.

Because some air-to-air radar operations and missile operations had been accomplished in both the earlier simulator flights, the rest of the flight concentrated on the air-to-ground mode of operations. A bridge near Chester, Mo., previously had been picked as an offset aiming point for a nearby target, and its coordinates had been entered into the inertial navigation system.

To demonstrate how the aircraft handled at low altitude and high speed, the throttles were advanced to military power until the aircraft reached 500 kt. At 5,000 ft., even in the 90F temperature and moderate turbulence, I was able to write on my knee board and push the appropriate controls on the cathode ray tubes to bring up the desired displays.

The aircraft's stabilator, ailerons and rudders were moving frequently to dampen turbulence at the low altitude. Throughout the flight, the digital flight control system was able to dampen any outside forces on the aircraft. Krings said the effectiveness of the flight control system could be noticed especially during gusty cross wind landings.

During the simulated low-level attack on a target, the radar display and the hands-on-throttle-and-stick (HOTAS) concept could be looked at more closely. With the right cathode ray tube in the air-to-ground delivery mode, the acquisition symbol on the display could be controlled from the acquisition button on the throttle. The workload to fly the aircraft, set the desired speed and move the acquisition symbol to its desired position was a little hectic at first, but after about 5 min. the acquisition symbol could be placed on the desired mode on the cathode ray tube relatively accurately and quickly.

While within the 80 naut. mi. range of the Hughes AN/APG-65 radar, the bridge was picked up as a hot spot on the radar and a diamond designator symbol was moved to the same location. As the designated target came closer, the radar automatically reduced its range scale.

During the approach, Krings said: "This is one of the few radars that can search for and define targets as you fly. With most others, you have to know where you are going before they are effective."

As the aircraft came within 20 naut. mi. of the bridge, the radar's range coverage automatically dropped to the corresponding range, and the expand Mode 2, which increases the scale of the target display, was picked by using the acquisition symbol and throttle controller to get a better look at the area around the bridge. From this range, buildings on both sides of the river could be seen on the radar, as well as barges in the river. At this time, the radar display also was showing on the front seat head-up display.

The head-up display and the left cathode ray tube in the rear seat were showing steering information to the offset target, while the radar was still pointed at the bridge. Also indicated on the display was a symbol that indicated whether radar, inertial or visual modes were being used to designate the target.

Closing on the target, a power plant next to water, the time-to-go counter decreased rapidly as the bomb line was approached. The simulated bombs were released automatically as the power plant was approached. At that point the target could have been redesignated and further attacks made from different quarters.

The radar then was put in the air-to-air search mode while at 5,000 ft. A target was picked up on the radar, which indicated that it was at an altitude of 8,000 ft., a speed of Mach 0.2 and traveling away and to the right of the F/A-18's flight path. Switching to the air-to-air mode did not necessitate the removal of hands from either the stick or the throttle. A British Aerospace AV-8A Harrier was picked up on the F/A-18's radar at 45 naut. mi., but because of fuel considerations, homing was not attempted.

The 70-naut.-mi. flight back to Lambert Field was utilized for trying the speed-brakes and getting the feel of the aircraft at lower speeds with the landing gear and flaps extended. Again, there was little, if any, pitch change during extensions or retractions of flaps, speed brakes and landing gear.

Use of Runway 30 at St. Louis offered the opportunity to set up a long final approach. Starting from a higher than normal altitude for the visual approach initially made it difficult to see over the front seat, and I had to look around the front seat for alignment. Once established on a 3-deg. glideslope, it was possible to look over the front seat to maintain alignment. The approach initially was made at 160 kt. to maintain separation from other traffic, then slowed to 130 kt. closer to the runway. Power was set to 84% engine rpm, and the angle of attack registered 8 deg. Throughout the approach, the aircraft was so stable I wanted to move the stick when it was not necessary, because I am used to flying aircraft without the digital fly-by-wire capability.

A slight increase in power was used to decrease the sink rate just prior to touchdown to smooth the landing. McDonnell Douglas pilots have found that while the F-15 has a tendency to float in ground effect, the F/A-18 maintains its glideslope angle to touchdown because of the carrier type approach used.

Krings took control of the aircraft and the nose wheel steering after landing and taxied the aircraft back to the blocks at the McDonnell Douglas facility. The Hornet was shut down with 800 lb. of fuel remaining after being airborne for slightly more than 1 hr., mostly in the high-speed maneuvering regime.

The single-seat F/A-18 has the benefit of 1,000 lb. more of usable fuel by the use of JP-5, rather than JP-4 on this flight, and a larger fuselage tank.

If any one segment of flying the F/A-18 can be called difficult for the uninitiated pilot, it would have to be the time required to learn all the modes and different symbology of the aircraft's avionics package, and especially the multitude of options in the navigation and weapons systems. Being accustomed to round dials, or steam gauges, as they are called, I had a little difficulty at first, but as time progressed I was able to absorb cathode ray tube display information more easily and less instruction was required.

McDonnell Douglas officials said all the pilots who have transitioned into the F/A-18 have had a little difficulty initially, but in a very short time navigation and weapon systems are run as if by second nature. This is especially true of the younger pilots who have grown up with computer keyboards and games.

Based on the flights in the simulators and the aircraft, the hands-on concept works in the F/A-18 and allows the pilot to keep his head out of the cockpit. Because of this, the head-up display is used primarily for flight information. Since there is no head-up display in the rear seat, I used the left cathode ray tube for flight information, and found no tendency to use the standby instruments.

Although the F/A-18 is not being promoted as an instrument weather attack aircraft, the capabilities observed on the short flight appear to give the aircraft that potential.

As for its carrier potential, the sea trials performed late last year were considered by the Navy and McDonnell Douglas to be the most successful in recent history. As a former naval aviator with more than 500 carrier landings—all in McDonnell Douglas A-4s—I believe the F/A-18, with its stability and relatively flat landing attitude, offering good visibility, would be an easy aircraft to bring on board ship.

Aside from the performance for the attack and fighter role, the F/A-18 has been pushed by the Navy as reliable and maintainable. Although that claim cannot be proved until the aircraft enters fleet service, preliminary figures indicate McDonnell Douglas is achieving the Navy's goals.

At the 1,200-hr. flight test point, raw figures for the mean flight hours between failures stood at 1.9. In this reliability evaluation all failures are considered equal, even though they could range from a hydraulic system failure to a light bulb failure. Correcting that figure for operational ground rules, the time between failures is 3.2 hr. To date, the F/A-18 fleet is achieving six maintenance man-hours per flight hour in organizational level unscheduled maintenance, while the design goal is eight maintenance man hours per flight hour.

F-15 fighter abilities evaluated

Robert R. Ropelewski/Eglin AFB, Fla.
April 26, 1982

Significantly enhanced air-to-ground capabilities of the Air Force/Mc-Donnell Douglas F-15 fighter with little apparent loss in its air-to-air capabilities are being demonstrated here in contractor-funded flight tests aimed at establishing the F-15's credentials as an effective dual-role fighter/interdiction aircraft.

Bulk of the systems integration and development testing associated with the $50-million McDonnell Douglas/Hughes Aircraft Co. effort to refine the air-to-ground capabilities of the F-15 has been completed, although additional testing and weapons qualification work is continuing while awaiting an Air Force flyoff of this aircraft, designated the F-15E, against the General Dynamics F-16E. The Air Force has expressed a requirement for up to 400 all-weather multipurpose strike fighters, and these are considered to be the two primary contenders for that role.

The competitive flyoff is expected to begin late this year or early in 1983 and could be finished by mid-1983. Flight testing of the F-16E by General Dynamics is scheduled to begin this summer. The F-15 Strike Eagle demonstrator, prototype for the F-15E, has been flying since September, 1980.

This *Aviation Week & Space Technology* pilot recently flew two flights totaling approximately 3 hr. in the F-15 Strike Eagle demonstrator to observe the aircraft in the air-to-ground environment, and an additional flight of approximately 1 hr. in an operational F-15D with the 555th Tactical Fighter Sqdn., 405th Tactical Training Wing at Luke AFB, Ariz., to observe the baseline fighter capabilities of the F-15. That flight involved multiple engagements against other F-15s examining different initial engagement scenarios.

Results of these flights, together with approximately 2 hr. of simulator work in St. Louis, suggest that the F-15 is approaching McDonnell Douglas' objectives to develop a fighter with night/all-weather strike capability and accuracies equivalent to those of dedicated attack aircraft in the day/visual delivery mode.

F-15 air-to-ground development efforts, referred to as the advanced fighter capability demonstrator program, were initiated in 1977 by McDonnell Aircraft Co. and Hughes Aircraft Co. (supplier of the F-15's APG-63 radar) to demonstrate enhanced F-15 air-to-ground attack capabilities while retaining established air superiority features of the aircraft. A key aspect of this initiative was the modification of the APG-63 radar to provide a high-resolution radar mapping capa-

113

bility using synthetic aperture radar technology. Using this technique, high-resolution mapping down to 8.5 ft. from a range of 10 naut. mi. was observed during our flights. This meant that the radar was able to distinguish objects and terrain features as small as 8.5 ft. and separated by as little as 8.5 ft. from other objects in the area.

The modified radar and other improvements were installed in the second two-place F-15B flight test aircraft, which has been leased back from the Air Force for this development program. Other modifications to the aircraft include:

- Replacement of rear cockpit instrumentation with four multifunction CRT displays, hand controllers and a digital electronic moving map to permit more efficient navigation and sensor control.
- An expansion of 50% in the memory capacity of the aircraft's central mission computer.
- Addition of a MIL-STD-1553 multiplex data bus to facilitate the integration of the new avionics systems and to communicate with the original avionics and data bus already in the aircraft.
- Higher rate of velocity and position updates from the Litton LN-31 inertial navigation system to the central mission computer. High rate of velocity updates, increased by 500% from the standard LN-31, is required by the synthetic aperture radar for high-resolution ground mapping. The more frequent position updates are critical to the central computer in the final stages of a weapons delivery run for high accuracy guidance to ground targets. There is no improvement in the overall navigational accuracy of the INS with these modifications.
- Integration of a Ford Aerospace & Communications Corp. Pave Tack forward-looking infrared imaging and laser ranging system. Evaluation of the Pave Tack pod on the F-15 is oriented toward a generic assessment of forward looking infrared and laser designator technology. An improved version of the advanced Flir pod being developed by Ford Aerospace for the Navy/McDonnell Douglas F/A-18 Hornet will be tested on the F-15 this summer. Program officials believe that the Lantirn Flir/laser system ultimately would be used if the Air Force continues development of that system.
- Addition of weapons stations to prototype F-15 conformal fuel tanks to increase the number of F-15 air-to-ground ordnance stations from three to five.

Five-station capability

Integration of these improvements has been a gradual process over the past year and a half. The initial demonstration of the five-station air-to-ground capability was accomplished in October, 1980, and high-resolution mapping tests were begun in November, 1980, and completed in August, 1981.

Flight testing of a Pave Tack pod on the aircraft did not begin until early this year, and my recent evaluation of the Strike Eagle represented only the fifth flight of the pod on the aircraft. Hand-off from the synthetic aperture radar to the Flir was performed during the first demonstration flight with impressive results. Program officials believe the precise target tracking and ranging made possible by the system will provide the F-15 with day/visual bombing accuracies under night/adverse weather conditions.

The first of my two flights in the Strike Eagle was dedicated to air-to-ground applications of the aircraft. Configuration for that mission included two conformal fuel tanks providing additional capacity for about 9,750 lb. of fuel, 12 500-lb Mk.82 general purpose bombs, four Raytheon AIM-9L Sidewinder air-to-air missiles, internal 20-mm. M61A1 Gatling gun, and the Pave Tack pod mounted under the fuselage centerline. The pod weight just under 1,300 lb.

McDonnell Douglas test pilot Gary L. Jennings, who has conducted most of the flight testing of the Strike Eagle, occupied the front cockpit for the flight while I became familiar with the controls for the avionics system and displays in the rear cockpit. Except for the radar controls, which are identical on the throttles in both the front and rear cockpits, the systems operator in the rear has command of all other sensor systems and displays. The front cockpit configuration in the demonstrator aircraft is the standard F-15 layout, although the pilot has the option of selecting radar imagery, Flir imagery or the electronic moving map display, called the tactical situation display, on the single CRT on the front panel.

The instrument panel full of CRTs in the rear cockpit included a 4-in. square screen at the extreme left and right of the panel with two 7-in. screens side by side in the center. The 4-in. display on the far left serves as a menu card for selecting the information to be displayed on the other screens. Radar, Flir, tactical situation display or head-up display symbology from the front cockpit HUD can be projected on any of the screens.

Jennings suggested using the left 7-in. screen for Flir imagery, the right 7-in. screen for radar imagery and the smaller 4-in. screen on the right for the tactical situation display initially, and I used the menu card on the left screen to set up the other displays. Two stick-grip type multiswitch control handles—one on the left console and one on the right in the rear cockpit—control the displays.

Takeoff weight

Our takeoff weight on this first flight was approximately 62,000 lb.— close to the F-15's maximum takeoff weight of 68,000 lb. Even so, the aircraft's high thrust-to-weight ratio provides an impressive takeoff performance.

Twelve seconds after Jennings released the brakes to start our takeoff roll, the aircraft was at 150 kt. and the nose was rotated to lift off at approximately 175 kt. after covering an estimated 3,800 ft.

The aircraft was leveled off at 5,000 ft. in a few seconds, and Jennings selected the air-ground master mode for the radar and central mission computer. Initial waypoint entered earlier into the mission computer was a tank hulk on a target range approximately 20 mi. north of Eglin Field. When the air-ground master mode was selected, the radar began automatically to search in the direction of the target range until it found the tank, which was at a bearing of about 45 deg. to our left as we flew northward. Using the radar controls on his throttle handles in the front cockpit, Jennings commanded a 1.3-naut.-mi. square patch map of the target area.

The synthetic aperture radar uses the movement of the aircraft with respect to the target to create the impression that the antenna is larger than it is. By scanning the selected area for several seconds—typically about 3-6 sec.—before producing a map, the radar is able to re-create the image of the surveyed area with finer detail than would be possible from an instantaneous picture. In this case, a 1.3-mi.

McDonnell Douglas F-15E Strike Eagle development aircraft, a modified F-15B, is shown taking off from Eglin AFB, Fla., for first of two Aviation Week & Space Technology *evaluation flights in the strike version of the F-15.*

F-15E Strike Eagle has conformal fuel tanks, Pave Tack target designator, 12 500-lb. Mk.82 bombs, and four AIM-9L Sidewinders.

square patch map provided a resolution of 17 ft. up to our range of 20 mi. from the target area. Beyond that range, and up to 50 naut. mi. away, it would have been necessary to call up a 3.3 × 3.3 naut.-mi. patch map with a resolution of 42 ft. Best resolution is within 10 mi., where a 0.67-mi. square patch map on the display provides a resolution of 8.5 ft.

The systems operator selects the area to be mapped by moving a cursor on the radar display with the help of a transducer switch—essentially a multidirectional "coolie hat" switch—on the top of the right control handle. A selected display window centered around the cursor moves with the cursor and can be placed over any smaller area of the radar display. Size of the display window can be adjusted.

Sequence initiation
Depressing one of the transducer switches on top of the right hand controller initiates the patch mapping sequence of the area within the display window. Patch map sizes available depend on the range from the area of interest, but they can be as small as 0.67 × 0.67 naut. mi. at 10 mi. to 80 × 80 mi. at 160 naut. mi.

In addition to the radar picture, digital readouts at the bottom and corners of the display provided aircraft heading, altitude and airspeed information as well as range and bearing to the position marked by the cursor, transmitter-on time remaining to complete a single look at the selected area, and maximum and minimum ranges covered by the display. A menu card along the bottom of the display allows the operator to select parameters such as map coverage and window size.

Although we continued to fly northward and did not conduct a simulated attack on the tank at that point, the radar continued to track the tank and the inertial navigation system continued to generate steering commands to the target on the head-up display until Jennings deselected the air-ground mode on the radar and central computer.

Turning my attention to the tactical situation display, which showed our position with respect to various landmarks, terrain features and navigational waypoints in the area, I moved the display's cursor over a spot designated Waypoint 6. This was Andalusia Airport, a small general aviation field in southern Alabama.

Digital readouts beneath the aircraft symbol near the bottom of the tactical display indicated that the airport was 28.5 naut. mi. away bear-

Four CRTs in rear cockpit of F-15E Strike Eagle provide aft crewmember with information from several sources. Outside screen on each side is 4 in. square, while two center displays are 7-in. screens. Screen on far left provides menu for selecting displays on other screens. Second screen from left in this case shows tactical situation display, an electronic map display on which several different scales can be selected. Large screen on right shows test pattern here, while the small screen at far right repeats head-up display imagery from front cockpit.

ing 36 deg. to the right of the F-15's nose. Using the switches on the right hand controller, I called up a small box representing the footprint of our radar and placed it over the Andalusia area. I then depressed the same switch used to move the footprint up, down and sideways, cueing the radar to begin looking in the direction of the airport.

We were above a low cloud deck at that point, still at 5,000 ft. In a few seconds, the radar display clearly showed the airport at a distance of 25 mi. At that range, it was possible to command a 3.3 × 3.3-mi. patch map with a 42-ft. resolution—clear enough to distinguish the intersecting runways of the airport and some buildings on the field.

Map resolution

Below 20 naut. mi. from the field, the radar was able to provide 1.3-mi. square patch maps with a 17-ft. resolution. Jennings selected the air-ground master mode once again on the mission computer and slaved the Pave Tack system to the radar, so that the infrared sensor was staring toward the same area as the radar. We were above the cloud cover and starting to descend into it, and the infrared system was still unable to see the Andalusia Airport through the clouds. The radar, inertial navigation system and tactical situation display provided a precise picture of our position with respect to the field.

Within 10 mi. of the airport, the radar was able to provide 0.67 × 0.67-naut.-mi. patch maps of the airport—an area within the airport's dimensions. Jennings had kept the airport about 45 deg. to the right of the aircraft as we approached—about the optimum angle for the synthetic aperture radar to achieve its best resolution.

As we reached about 5 naut. mi. from the field, we froze the radar to stop any further signal emissions, and Jennings put the aircraft in a descending right turn toward the field for a simulated attack. The INS and central mission computer continued to provide steering cues for the attack.

About this time we began to emerge from the bottom of the cloud layer, allowing the Pave Tack system to provide clear imagery of the airport on the left primary (7-in.) display screen. We acquired the field visually at about the same time, and shortly after this the bomb release cue began flashing on the head-up display, signaling that our bombs would have released automatically at that point had the weapons system been armed.

We turned away from the field and began climbing as Jennings depressed the mode reject switch in the front cockpit to take the mission computer out of the air-ground master mode and return it to the air-to-air mode. The Pave Tack infrared sensor remained locked on to the runways at Andalusia, directly behind us. In an actual combat situation, this would have permitted a preliminary damage assessment without the need for an aircraft to stay in the area for that purpose.

After several minutes I squeezed the trigger on the front of the right hand controller to return the Pave Tack sensor to the forward boresight mode. A small white diamond-shaped symbol on the Pave Tack display showed where the system was pointing at any given moment with respect to the longitudinal and directional axes of the aircraft.

Manual control

In addition to slaving the system to the radar, it also can be controlled manually to scan any area. Once a potential target is acquired, the infrared sensor can be locked on so that it is no longer necessary to track the target manually.

As we climbed away from Andalusia on a northwesterly heading, I moved the cursor and the radar footprint on the tactical situation display to our next target—a bridge and dam at Miller's Ferry, about 50 mi. away on the Alabama River. The radar then was cued from the tactical situation display, and in a few seconds it produced a terrain map centered on the bridge and dam. Digital readouts on the radar display indicated the target area was 42 deg. to the left of our heading at a range of 54.5 mi.

Moving the cursor and selected display window over the target on the radar screen, I called up a 4.7 × 4.7-mi. patch map of the area. The bridge, dam, roads leading to the bridge and other landscape features were clearly distinguishable when the first map appeared 6-7 sec. later. Jennings had leveled the aircraft at 15,000 ft., resulting in a radar grazing angle of about 2.5 deg. between the aircraft and the targets. Flight tests have shown that the Hughes radar is capable of providing high-resolution mapping down to grazing angles as low as 0.5 deg. using synthetic aperture techniques.

Some sunlight was falling on the displays in the rear cockpit at this point, but the imagery remained readable. "We're not totally free of sunlight problems yet," Jennings said, "but it's a lot better than it used to be and we're pretty satisfied with it."

At a true airspeed of 380 kt. and a ground speed of 340 kt., the aircraft was closing the distance quickly. There was a steady stream of commands to be made to the radar, Pave Tack and other systems to prepare for a simulated attack on the dam. When our range dropped below 50 mi., we went from 4.7-mi. patch maps to 3.3-mi. maps on the radar display. At 27-28 mi., we slaved the Pave Tack Flir to the radar's line of sight, causing the Flir to acquire the dam and bridge. There was no cloud cover in this area, and we quickly had a crisp image of the area on the Pave Tack display.

Detailed imagery

At 20 mi., fine details were apparent in the radar imagery of the dam and bridge, and Jennings began a descent for a simulated low-level attack while I switched from 3.3- to 1.3-mi. square patch maps on the radar.

We kept the dam about 20 deg. to the left of the aircraft's nose until we had closed to within 10 mi. At that point I called up a 0.67- × 0.67-mi. radar patch map centered on the dam, then froze the radar to stop any further emissions.

From 10 mi., the radar imagery was very detailed, with individual pillars and abutments discernible on the dam. On his radar display, Jennings designated the east end of the dam for our attack, cueing the weapons delivery computer and HUD symbology to provide guidance to that point. We turned toward the dam, leveling off at about 1,000 ft. In an actual combat situation, the altitude undoubtedly would have been much lower.

Although we intended to make a straight pass across the dam in this case, the Strike Eagle weapons delivery system has provisions for several different delivery options, both automatic and manual. A loft delivery from a standoff position would have been possible in this case, for example. Using the Pave Tack system for laser designation and ranging, it also would have been possible to use such standoff weapons as the GBU-12 laser-guided bomb.

Once the location of the target is updated in the mission computer from a radar fix on either the target or an offset aim point, command steering symbology is projected on the head-up display for the pilot to follow, and weapon release is automatic.

McDonnell Douglas is working on a maneuvering delivery mode that will allow pilots to maneuver continuously during their weapons delivery runs while the mission computer keeps track of the aircraft's position with respect to the target and determines when to release the weapons.

In this case, we made a straight and level run across the east end of the dam, following the steering commands on the head-up display. The HUD flashed and an audio tone was heard signifying that our weapons had been released automatically as we approached the dam. Jennings turned hard left to depart the area heading back toward the southeast, and I took the controls.

Staying at approximately 1,500 ft. altitude, I advanced the throttles to military power—the maximum power setting before lighting the afterburners. Our speed increased and eventually stabilized at around 530 kt.

Ride quality
In the past, there have been complaints from operational F-15 units about the ride quality of the aircraft. On this occasion, the ride was comfortable—probably due to the heavy bomb load we were still carrying under the wings. There was light turbulence in the area, and the aircraft occasionally exhibited a slight jittering, particularly when we passed close to some of the small cloud buildups at our altitude.

The aircraft was limited to 5g with the external load we were carrying, and that limit was reached very quickly as I rolled the aircraft into a tight left turn. Roll rate was excellent despite the heavy underwing loads, and there was little energy loss in sustained 5g turns under those conditions. A series of windup turns left and right caused the airspeed to drop to about 480 kt., but it climbed quickly back to 530 kt. when the maneuvering stopped.

Without the heavy air-to-ground ordnance, the aircraft would have been capable of sustained turns at 7g and above without any loss of airspeed or altitude. This was observed first-hand during two later flights in the Strike Eagle aircraft and an F-15D.

We continued to exercise the radar and Pave Tack displays during this time and at one point lost information from the configuration menu card on the left 4-in. CRT as well as information from the tactical situation display, which at that point was on the left primary (7-in.) screen. The same incident occurred near the end of the flight, and both times the displays were restored by recycling the mission computer.

Another difficulty also occurred around this time when the radar, which had just begun to paint Eglin AFB about 40 mi. away, began to provide imagery with very heavy shadowing across much of the picture. Jennings explained that this was due to the strong return signals from large hangars and other buildings on the base. This caused the radar logic to turn down receiver gains, reducing receiver sensitivity below the level needed to provide an adequate picture of areas with low radar reflectivity. These areas appeared very dark on the cockpit display. Fixes are being sought for this, Jennings said.

En route to Eglin, we executed another simulated attack on Andalusia Airport, this time catching a twin-engine business aircraft taxiing down one of the runways and tracking it with the Pave Tack Flir. Heat from the taxiing aircraft's engines was clearly discernible through the Flir, and it was easy to keep the Pave Tack cross hairs centered on either one of the two engines.

Continuing back to Eglin, I slowed the aircraft and gradually raised the nose until the aircraft was stabilized in level flight at 35 deg. angle of attack. Keeping the angle of attack between 30 and 35 deg.—an uncomfortable and risky attitude in many other aircraft —I executed a series of turns that brought out no unusual characteristics, no buffet and no other signs of distress in the F-15.

We landed at Eglin 1 hr. 48 min. after takeoff, and McDonnell Douglas crews removed the air-to-ground ordnance and Pave Tack pod from the aircraft in a demonstration of the relatively quick and easy convertibility of the Strike Eagle from an all-weather strike aircraft to a fighter configuration. Approximately 1.5 hr. later, we were airborne again armed only with the 20-mm. Gatling gun and four AIM-9L Sidewinder air-to-air missiles.

Our second flight in the Strike Eagle was intended to demonstrate that the F-15 lost little of its effectiveness as an air-to-air weapon platform with the incorporation of the enhanced ground attack capabilities. From the moment the throttles were advanced for takeoff, there was little doubt about this. Less than 1 min. after releasing the brakes, we were leveling the aircraft at 10,000 ft. not far beyond the upwind end of the runway and turning northward again to enter the Eglin range area.

Prior to doing our air-to-air work, we took the time to run two simulated bombing runs at low altitude on the tank target we had only looked at on the radar at the beginning of our earlier flight. Range scheduling requirements had prevented us from pressing the attack on the first flight. I flew the first pass against the tank while Jennings flew the second run.

Tank acquisition
The radar acquired the tank almost immediately from our range of about 20 mi., and I began a descent to approach the target at low altitude. Flying a 30-deg. right offset approach toward the tank, we took our last radar fix at approximately 7 mi. from the target before freezing the radar, then followed the steering commands on the head-up display for the final run.

Following the run-in line on the HUD resulted in the aircraft's passing about 150 ft. to the left of the tank rather than directly over it. From a weapons delivery point of view, this was unsuitable. Jennings flew the next pass after returning to a point beyond 20 mi. from the target to start the run. As on the first pass, the steering cues brought the aircraft to a point 100 ft. or so to the left of the tank, and Jennings surmised that the inertial navigation/weapon delivery system—one of the earliest LN-31 units produced by Litton—may have needed calibration.

One problem that I experienced in this case and at other times when designating targets on the radar, Pave Tack display and tactical situation display was the type of switch used to accomplish this.

Cursor position
Depending on the display in question, a transducer switch on the top of either the left or right hand controller is used to position a cursor over the target to be designated. This is done by pressing forward, aft, left and right on the switch.

Once the cursor is over the target, the same transducer switch is then used to designate the target for the radar or Pave Tack. Theoretically, this use of the same switch to move the cursor and designate the target saves space and improves operator efficiency. The problem, however, was that depressing the transducer switches to designate targets required absolute discipline to ensure that the switches were pressed straight down. If they were not, unintentional directional inputs were fed to the cursor, moving it off the intended target by a small amount on the display.

We departed the target range at that point and climbed initially to 10,000 ft. to exercise air-to-air capabilities of the Strike Eagle. With military power applied, I rolled the aircraft into a 3g left turn with a starting airspeed of 350 kt. Airspeed slowly decayed and finally stabilized at 250 kt., with the aircraft just beginning to buffet slightly.

Lighting the afterburners, we accelerated to 400 kt. and entered another sustained turn, this one at 5-6g, while maintaining airspeed at 400 kt. Rolling the wings level, I brought the throttles to idle, brought back the stick to acquire 30 units angle of attack, then held that through an idle-power loop. Aircraft weight at that point was approximately 44,000 lb.

Among the most impressive aspects of the F-15 is its controllability, power on or power off. In this first of several loops, the aircraft's speed over the top of the loop dropped to 60 kt. because I held slightly too much back pressure on the stick. Yet there was no difficulty in controlling the aircraft and continuing through the rest of the maneuver.

Subsequent loops were initiated with military power and then with full afterburner, starting at 400 kt. and slowing to about 150 kt. at the top in both cases.

Angle of attack

We climbed to 15,000 ft. briefly, brought the power back slightly and began increasing our angle of attack while maintaining altitude. A slight buffeting began at 35 deg. angle of attack, and I began adding power while continuing to raise the nose. The aircraft eventually reached 47 deg. angle of attack while maintaining its altitude.

After performing several other maneuvers, we had descended to 8,000 ft., where Jennings suggested pulling the throttles to idle and raising the nose 60 deg. until airspeed dropped to 250 kt. When we reached that point, I pulled the nose up and over to complete the remainder of a deformed loop in which the airspeed indicator dropped to zero at the top. Despite this, enough elevator authority remained to enable me to continue pulling the nose downward and finish the rest of the maneuver.

Afterburners ignited

We leveled off at 6,000 ft. with an airspeed of about 400 kt., and I ignited the afterburners and rolled into a right turn pulling 6.5g—the maximum allowable at the aircraft's weight then. The F-15 accelerated to 600 kt.—an example of the sheer force of this aircraft with its two Pratt & Whitney F100 engines—before I brought the throttles out of afterburner and turned toward Eglin to end the flight.

Program officials at McDonnell Douglas said that development work is essentially finished on the sensors, controls and displays for the air-to-ground enhancement of the F-15, and the remainder of the demonstration program will concentrate on weapons deliveries with a wide range of air-to-ground weapons. This will include most types of general purpose ordnance, guided weapons, munitions and dispensers, and a 30-mm. gun pod. Most of these have been qualified on the F-15 in earlier tests.

Air Force crews are scheduled to conduct a series of 10 flights in the Strike Eagle aircraft in May and June for a generic evaluation of the synthetic aperture radar and Flir/laser technology incorporated in the aircraft. The assessments made by the Air Force crews during these flights could be applied later to the anticipated F-15E/F-16E flyoff.

F-16 displays combat capabilities

Robert R. Ropelewski/Carswell AFB, Texas
May 28, 1979

Simulated engagements at various altitudes demonstrate increased g-tolerance, versatile HUD during evaluation flight.

High thrust, excellent maneuverability, increased g-tolerance and a versatile head-up fire-control system have produced a significant improvement in close-in air-to-air combat capabilities with the USAF/General Dynamics F-16 lightweight fighter-capabilities that were underscored during a 1.3-hr. evaluation flight by this *Aviation Week & Space Technology* pilot.

The *Aviation Week* flight, first in the F-16 by other than test and military pilots or government officials, included air-to-air engagements against other aircraft at higher and lower altitudes as well as a brief look at the F-16's air-to-ground capabilities.

U.S. and European air force pilots currently transitioning into the F-16 from such aircraft as the McDonnell Douglas F-4 have been impressed by the performance of the new aircraft, describing the transition as analogous to the change from propeller-driven to turbojet fighters in the late 1940s and early 1950s. On the basis of this pilot's background as a former Marine Corps aviator and more recent experience in numerous high-performance aircraft around the world, the *Aviation Week* flight provided a convincing display of the F-16's dogfighting capabilities.

The F-16B is a two-seat version of the General Dynamics F-16A, and is powered by the Pratt & Whitney F100-PW-100 turbofan engine.

The aircraft flown was the 10th twinseat F-16B to be produced and the 17th worldwide production aircraft in the F-16 series. The aircraft had just returned from a one-week demonstration tour in Canada and was scheduled for delivery to USAF at the end of May.

Neil R. Anderson, director of flight test for General Dynamics and chief test pilot on the F-16 program, accompanied this pilot on the flight. Because an engine emergency air start capability exists only in the front seat, Anderson occupied the forward seat and this pilot flew from the aft seat.

The high thrust-to-weight ratio made possible by the F-16's single 25,000-lb. thrust Pratt & Whitney F100-PW-100 turbofan engine and performance were evident as soon as the throttle was advanced for takeoff from the runway at Carswell, where General Dynamics' Ft. Worth plant is located. Weight of the aircraft at the start of the takeoff roll was approximately 22,500 lb., including about 5,900 lb. of fuel and two wing-tip mounted AIM-9L Sidewinder missiles. This was close to the F-16's normal air-to-air mission weight of 23,000 lb. Field elevation at Carswell was 640 ft., and the temperature was 75F with a 10-kt. headwind.

At the 22,500-lb. takeoff weight, thrust exceeded aircraft weight and the acceleration was like a kick, even before the throttle had passed from military power into afterburner. Ten seconds after the roll began, airspeed was passing through 125 kt. and increasing rapidly. Anderson, flying the aircraft for this initial takeoff, rotated the nose upward, stopping at 60-deg. pitch as the aircraft began climbing out after a ground roll of approximately 1,200 ft.

Acceleration continued, even in that attitude with the aircraft passing through 170 kt. about 30 sec. after brake release. A wingover maneuver was used to level the aircraft at 8,000-ft. altitude, still within the length of the Carswell runway. A USAF/Northrop T-38 chase aircraft, which had started its takeoff roll on the same runway 5 sec. after the F-16, was just lifting off the runway below.

Anderson noted that the aircraft could have climbed out at a 90-deg. pitch attitude after liftoff. However, the pilot's seats are tilted aft 30 deg. for better load factor (g) tolerance, and that plus the 60-deg. pitchup put us literally flat on our backs. A higher nose attitude would have placed our heads downward in a negative-g position.

Shortly after leveling off from the takeoff climb, Anderson turned over controls of the F-16 to this pilot for most of the remainder of the flight. We were heading toward a reserved block of airspace north of Ft. Worth, where, after some basic performance and handling qualities exercises, we would engage in air-to-air maneuvering with other aircraft.

Apart from a slightly reduced internal fuel capacity, the F-16B has the same general dimensions, weights and performance and handling characteristics as the single-seat F-16A. General Dynamics pilots and engineers stress that it can be used in combat as effectively as the single-seat aircraft.

Cockpit of the F-16 is uncramped and arranged for one-man operation of all systems. The objective has been to maximize head-up flying and systems operation. Outside visibility is excellent from the F-16 cockpit because there are no structural members forward or above the pilot except for the canopy itself. The full canopy eliminates the need for a separate windscreen, thereby eliminating the need for structural framework directly in front of the pilot.

A fairing behind the canopy of the F-16B limits rearward visibility somewhat for the rear seat occupant, but the pilot in the front seat of the F-16B or in the single seat F-16A also has essentially unrestricted visibility over his shoulders as well. The bulbous canopy of the F-16 also permits excellent downward visibility, and offers the possibility of looking straight downward by banking the aircraft very slightly.

Primary difference from other aircraft cockpits is the side-stick controller on the right console, replacing the centermounted control stick found on most aircraft. Absence of the stick in front of the seat allows the between-the-knees space there to be used for an extended console housing the radar/electro-optical display, angle-of-attack vertical tape display and horizontal situation indicator. Just above the last two indicators, at the bottom of the center instrument panel, are the airspeed/Mach indicator and attitude indicator.

Toward the top of the center instrument panel is the head-up display control box, with the head-up display (HUD) itself mounted atop the control box. The Marconi-Elliott HUD is an adaptation of the system used in the LTV A-7, and it provides target identification and acquisition, gunnery and missile solutions, weapon aiming and flight reference and energy management cues.

A radar threat warning indicator and weapons management box fill the panel to the left of the HUD controls, and a standby attitude indicator, barometric instruments and engine and fuel gauges occupy the right side of the front panel.

An adjustable arm rest is mounted on the right cockpit wall for use with the side-stick controller, and a wrist rest can be extended from the wall several inches ahead of the arm rest. The arm rest is essential in high-g flight, but the wrist rest is optional and was not used on this flight.

It takes only a few minutes to become accustomed to the position of the stick, and to the fact that it is force sensing and does not move.

The stick is sensitive without being troublesome and there was no tendency toward pilot-induced oscillations, even in the first few minutes of getting accustomed to it. To perform a gentle roll to the left or right required hardly more than the thought. A firm push to the left or right produced an instantaneous high roll rate-up to 240-deg. per second at speeds of 250 kt. and above.

Likewise, pitch response depends on the force applied to the stick rather than its actual movement. The response is immediate. An abrupt tug on the stick produces an immediate g load corresponding to the force applied to the stick.

In both roll and pitch, there was no need to check the rolling or pitching by pushing the stick in the opposite direction. Simply releasing pressure on the stick brought movement about the roll or pitch axis to an immediate halt.

Pilots who have flown in high performance two-seat aircraft are accustomed to knocking their hardhats against the canopy when the other pilot snaps the aircraft into a roll. Stopping the roll seldom presents the same problem.

General Dynamics F-16 with landing gear extended and smoke generators in the wing tips flies at an air show.

In the F-16, however, the abrupt halt caused by simply releasing pressure on the stick is enough to straighten the second pilot's hard-hat on the other side of the canopy.

Like the stick, the rudder pedals have no perceptible movement (about ½ in., in reality). The pedals sense force applied to them and cause a yaw displacement in the aircraft based on the amount of force applied.

The nature of the fly-by-wire controls in the F-16 allows them to be tailored through the computer to provide whatever performance characteristics are desired. General Dynamics engineers have taken advantage of this in several ways, including the imposition of some limits aimed at preventing pilots from losing control of the aircraft.

The main limits are in the pitch axis, where a load factor/angle-of-attack limiter has been incorporated, and in the yaw axis, where a rudder fadeout feature has been incorporated to limit pilot rudder pedal command when the angle of attack goes above 20 deg. In addition, an aileron-rudder interconnect and roll-rate-to rudder feedback reduce coupling between the roll and yaw axes.

All of these features are intended to make the F-16 a head-up aircraft in which the pilot can devote full attention to events outside the cockpit without risking loss of control of the aircraft.

The load factor/angle-of-attack limiter in the pitch axis prevents the pilot from pulling more than 9g, or exceeding 25.3 deg. angle of attack, no matter how hard he pulls on the stick. At high speeds, the 9g limit is reached with 31 lb. of back force on the stick. Any more force than this will still produce only 9g.

These limits were tested initially on the *Aviation Week* flight in an air-to-air encounter with the T-38 chase plane. The T-38 was 2-3 min. behind the F-16 in arriving in the maneuvering area north of Ft. Worth.

As we circled in the F-16 to await the other aircraft, the T-38 passed by a few thousand feet below in the opposite direction from the F-16 heading. This pilot rolled the F-16 into a descending left turn and applied military power to intercept the T-38. Crew of the T-38 spotted us at that moment and began a tight left turn in afterburner.

Speed of the F-16 at the start of the maneuver was too low to reach 9g, even with maximum back pressure on the stick. However, the 25-deg. angle-of-attack limit was reached quickly, and the limiter held the aircraft there. Turn rate of the F-16 was still extraordinary, and the aircraft took only a few seconds to establish itself in a tracking position close behind the T-38.

Forward visibility is unrestricted in the F-16A and in the front seat of the F-16B, making it easy to follow every move of an aircraft ahead. With no windscreen framework to obstruct the view, all that remains above the instrument panel is the head-up display see-through lens with a small angle-of-attack indicator to the left of it and a small radial refueling/nose wheel steering mode indicator to the right. Standard wet compass normally found at the top of the windscreen frame of most aircraft has been installed at the lower right corner of the front instrument panel.

Visibility from the rear seat of the F-16B is restricted by the other pilot's seat when looking directly forward, but by leaning the head to one side or another and staying inside the turn of the T-38, this pilot could continue tracking the other aircraft from the back seat. Unlike the front, the rear cockpit is not equipped with a head-up display.

Afterburner was lit to break away from the T-38 and the F-16 accelerated quickly from 250 kt. to over 400 kt. Altitude then was 15,000 ft.

Shortly after this, the F-16 and T-38 were cleared into a reserved block of airspace between 27,000 ft. and 41,000 ft. for further air-to-air maneuvering. In a level turn in military power at 27,000 ft., the aircraft developed a mild shudder for a second or two, described by Anderson as "Mach dither," as the aircraft passed Mach 0.95 entering the transonic flight regime. It was the only time in the flight that this pilot detected any sign of the passage from subsonic to supersonic or vice versa.

Throttle was pulled to idle at this point, the nose was raised to about 50-deg. pitch and the aircraft immediately reached its angle-of-attack limit of 25 deg. The power-off climb peaked at 36,000 ft., at an indicated airspeed of 103 kt. Several firm tugs on the stick failed to increase the angle of attack beyond 25 deg., and the aircraft began sinking slowly but would not stall. A very light buffeting could be detected, but there was no rock or other control anomalies. Ailerons remained effective, though less responsive because of the low speed.

Because the angle of attack was above 20 deg., the rudder fade-out feature was acting to limit available rudder command and thus prevent a spin. Only a minimal yaw could be induced, no matter how hard the rudder pedals were pressed. The aircraft continued to sink slowly forward.

We pushed the nose down steeply and lit the afterburner to accelerate to supersonic speed. There was no shuddering, pitching or trim change to indicate the passage through the transonic regime.

The F-16 has a top speed of Mach 2, but because there are no unusual characteristics to observe at that speed, the acceleration was stopped at Mach 1.38. The throttle was brought out of afterburner to military power, and a tight left 4-5g turn was entered to decelerate to subsonic speed. No Mach tuck, bumps or pulsations in the flight controls appeared as the aircraft's speed dropped below Mach 1.

Still turning at 33,000 ft., back pressure on the stick was increased to get maximum g. Speed was bleeding off too quickly at that point however, and the best that could be achieved was 6.5 g before reaching the angle of attack limit.

One of the reasons for the impressive turning performance of the F-16 is the automatic wing cambering and contouring that takes place as the aircraft moves through the flight envelope. Four different wing contours are programmed automatically depending on the maneuver. These configurations include:

- **Takeoff and landing ground roll**—Wing leading edge is 2-deg. above its centered position, and the trailing edge flaps are deflected 20-deg. downward.
- **Takeoff climb and approach**—Leading edge is deflected downward 15-deg. and trailing edge is deflected downward 20-deg. after wheels leave the ground on takeoff and when landing gear is extended during the landing approach.
- **High-speed cruise**—Wing leading edge and trailing edge are both positioned 2-deg. over center.
- **Maximum maneuver**—Leading edge is drooped 25 deg. and trailing edge is centered.

The pilot can watch the leading edge change from one configuration to the next as speed and flight circumstances change.

Also contributing to the turn performance and directional stability of the F-16 are the forebody strakes that extend along either side of the fuselage from the wing leading edge to the front of the cockpit. The strakes generate strong vortices at high angles of attack, improving air flow over the wing and empennage. The result of this, according to the F-16's designers, is increased lift, reduced buffet and better directional stability at high angles of attack. When temperature and dew point are close together, as was the case during this particular flight, the strakes can be observed doing their work by the condensation trails that swirl across the wing in the vortices.

The combined effects of the strakes and the automatic wing cambering reduce buffeting to almost imperceptible levels in most phases of flight in the F-16, although Anderson conceded that moderate buffeting can be felt in hard maneuvering at supersonic speeds. On this flight, the level of buffeting never reached an intensity where it adversely affected target tracking in air-to-air maneuvering.

Following the 6.5g turn at 33,000 ft., this pilot pulled the throttle to idle and raised the nose to 45-50-deg. pitch attitude. At 150 kt., the aircraft was rolled inverted and the nose was pulled through to execute a split "S" maneuver riding the 25-deg. angle of attack limit throughout. Airspeed remained at 150 kt. through the recovery, which used about 3,000 ft. of altitude.

After another one of these maneuvers, the aircraft was at 27,000 ft. and the throttle was pushed into the afterburner range to accelerate and regain some altitude. As the throttle was pushed forward, however, there was a loud thump and the afterburner was immediately deselected.

The thump was an afterburner blowout caused by the low speed and high altitude of the aircraft when the afterburner was selected. The F100 engine is equipped with a five-stage afterburner. First two stages had ignited properly, but there was too little airflow for subsequent stages and the raw fuel from the third stage then extinguished the first two stages. This pilot set the throttle at military power and pushed the nose down to accelerate to a higher airspeed before trying the afterburner again.

After climbing back up to 31,000 ft., the F-16 was turned for a head-on intercept of the T-38 that had accompanied us on takeoff. The T-38, flown by General Dynamics test pilots Dave Palmer and Alex Wolfe, was at that moment about 80 mi. to the south at an altitude of 38,000 ft. The objective was to track the T-38 inbound for an intercept using the look-up air-to-air capabilities of the F-16's Westinghouse X-band pulsed Doppler radar.

A Delco general purpose digital fire control computer with a 32,000-word memory is the key to the F-16 weapons package, managing the activities of the radar, radar/electrooptical displays, head-up display, inertial navigation system and stores management system. On this flight, the computer failed about 2 min. after takeoff. This meant that while the other system could operate independently, there was no computational capability to provide us with lead computation on the head-up display, for example, or with target speed and altitude information on the radar display.

No further problems

The computer was turned off for several minutes, and then turned on again a short time later. It worked—without problems—for the rest of the flight.

In an air-to-air combat situation, the pilot can operate just about everything from either the throttle or the stick. In the case of the radar, with the air-to-air mode selected on the radar control panel in the front seat, the pilot in the aft cockpit can take control of the radar by pushing a button on the left side of the throttle. The same button then serves to move a cursor-two short parallel vertical lines-on the radar scope.

Targets are presented as small squares on the scope. Any one of the targets can be isolated by moving the cursor so that the small square representing the target is between the two parallel vertical lines. The target is then designated by pushing upward on a designate/search button on the side-stick controller.

Once the target is designated all other targets are removed from the display. The small square turns to a diamond, the cursor lines disappear and range, speed, g-load altitude and closure rate information on the target appear on the radar display. The radar can be returned to the search mode by pushing downward on the designate/search button on the stick.

After the pilot has designated the target, a box appears on the head-up display showing the pilot the area in which the target aircraft is operating. He can, at the same time, select the weapons mode.

The basic air-to-air weapons of the F-16 are two AIM-9 Sidewinder infrared heatseeking missiles and a 20-mm. gun. Space also has been reserved for the systems necessary to fire radar-guided missiles from the F-16 in the future.

The missile firing mode can be selected either on the stores control panel on the front instrument panel, or with a dogfight/missile override switch on the top of the throttle. The HUD then displays dynamic missile launch zone information based on the movements of the target.

Two air-to-air gunnery modes are selectable from the stores control panel—a snapshoot mode providing electronic tracers for transient firing opportunities, and a lead computing optical sight mode providing a lead-angle computation for encounters where smooth angle tracking of the target is possible.

A dogfight mode, selectable by a switch on the throttle, overrides all other modes and orients the entire fire control system to an air-to-air mode. This feature is important when the aircraft is engaged in air-to-ground operations and must switch quickly to air-to-air status. When this mode is selected, the radar automatically searches a 20 × 20-deg. field of view of the HUD and locks on the closest target within 5 naut. mi. Target range, angle and rates of change are fed to the fire control computer. Target position, dynamic missile launch zone and snapshoot gunnery solutions are then displayed on the HUD.

Radar controls in front

It was possible to operate and control the radar from the back seat of the F-16B, as far as selecting scan elevations and designating and tracking targets. But only the front seat occupant has a complete radar control panel, located just beside the throttle, which permits selection of air-to-air and air-to-ground modes, range scales and scan width.

The same radar control panel also allows the pilot to select any of four slightly different frequencies so that when several F-16s are flying together, their radar emissions do not interfere with one another.

On this flight, the radar had been painting numerous unknown targets—probably transport aircraft—at some impressive distances. However, although the general direction, range and altitude of the T-38 were known, the F-16 radar was unable to detect the other aircraft. The radar did appear to pick up the T-38 briefly at a distance of about 40 mi., but when an attempt was made to designate the target for tracking, it was lost from radar display. Even as the T-38 came into visual range, the radar did not show it.

As the other aircraft passed in the opposite direction at 38,000 ft. and began a turn to the right, this pilot rolled the F-16 into a climbing right turn in military power that quickly positioned the aircraft inside and behind the T-38. The radar still did not pick up the target. After the flight, ground technicians said one of the mode switches on the radar control panel was found in the wrong position, possibly causing the radar to be ineffective in the look-up mode.

After breaking off from the T-38, we descended in the F-16 to 7,000 ft. for a low altitude encounter with an Air Force Reserve/Fairchild Republic F-105 flown by Capt. Gilbert D. Mook. During the descent, several turns were executed in which the F-16 reached a load factor of 8.7g-still slightly below the aircraft's 9g limit.

The 30-deg. inclined seats, which are sunk slightly into the floor of the cockpit, emphasize the pilots' semiprone positions and are extremely effective in raising the g tolerance in the F-16 cockpit. There was no indication of gray-out or tunnel vision during any of the maneuvers at 8g or more despite this pilot's past experience with the onset of tunnel vision at 6.5-7.0g.

One aspect noted several times during the flight was the tendency to pull back as hard as possible on the stick after the angle-of-attack or g limits had been reached in an attempt to get still more turn rate from the aircraft. Air Force pilots have noticed this tendency, and attribute it to the fact that the stick has no physical stops, since it does not move, to signal the pilot he is at the limit.

For this reason future F-16s will have a side-stick with very slight movement capability-about one-eighth of an inch-that will give pilots a hard limit they can feel when they have reached full control deflection.

As was agreed the F-105 began a run-in at 1,000 ft. on a course opposite that of the F-16. The F-16 radar worked well in the look-down mode, and the F-105 was identified and designated while still several minutes away. The F-16 was flown so that the target was slightly to the left side on the radar screen. Data block for the aircraft showed it at 1,000 ft. traveling at 450 kt.

As the two aircraft closed and the F-105 passed under the left wing of the F-16, it broke hard left and began accelerating in after-burner in a climbing turn. This pilot rolled the F-16 into a hard left descending turn at just over 8g in military power.

The excellent visibility of the F-16 made it possible for Anderson to watch the F-105 continuously from the front cockpit, but there was a brief period when this pilot could not keep the other aircraft in sight from the rear cockpit. This was only momentary, however, and the F-105 was acquired again from the rear cockpit after about 90-deg. of turn. After 180-deg. of turn in the F-16, the F-105 had turned about 135-deg., and it was apparent that the target aircraft could not outmaneuver the F-16.

After 270-deg. of turn in the F-16, the conversion was all but completed. With a few more degrees of turn, the F-105 was within the missile firing envelope of the F-16, and with a few more degrees was within gun-firing range.

A brief look at the F-16's air-to-ground capabilities showed the aircraft and its fire control system to be effective in that mode as well. The expanded and Doppler beam sharpening features of the radar in the ground mapping mode provided excellent images of potential ground targets, including distinct views of runways, taxiways and aircraft at Carswell AFB several miles away.

Returning to Carswell, Anderson demonstrated a simulated flame-out approach and a touch-and-go landing before this pilot made the final landing. The aircraft is as easy to handle in the landing pattern as elsewhere in its flight envelope even from the back seat.

Sensitivity of the flight controls is cut by approximately half when the landing gear is lowered so that overcorrections are minimized. Wing leading edge droops 15-deg. and the trailing edge 20-deg. when the landing gear is lowered. The aircraft has an angle of attack index on top of the glare shield and another on the instrument panel, and either one can be used to fly the aircraft to a landing at 13-deg. angle of attack. This gives the F-16 an approach speed of about 125 kt. at normal landing weights, and a touchdown speed of 118-120 kt.

Visibility from the back seat of the F-16B is adequate to land the aircraft, even though there is no visibility directly forward. The bubble canopy permitted this pilot to lean his head far enough to one side or the other to view the runway adequately.

The landing was the first point in the flight where it became necessary or useful to trim the F-16. Until that point, the fly-by-wire control system and the variable camber wings had automatically canceled out any changes in pitching moment, obviating the need for pitch changes in the cockpit.

Even the landings could have been flown easily without any trim changes. However, full nose-up trim sets the aircraft at 13-deg. angle-of-attack, which is the reference attitude for the approach. This left nothing to do but align the aircraft with the runway and adjust power to control rate of descent.

Once on the ground, the F-16's heavy duty brakes were used to stop the aircraft in slightly less than 2,000 ft.

On the ground, the aircraft taxied easily with and without nosewheel steering. Without nosewheel steering, the aircraft could be steered comfortably with differential braking, although this required an occasional increase in power to keep the aircraft rolling. Nosewheel steering was engaged by pushing a button on the front of the stick. A small indicator light with the letters NWS illuminated beside the head-up display to confirm that nosewheel steering was engaged.

Fifteen production F-16s have been delivered to the U.S. Air Force so far, with two others to the Belgian air force. The Dutch air force will receive its first production F-16 in early June.

Chinese FT-6s
enhance Pakistan fleet

John E. Fricker/Mianwali Air Base, Pakistan
April 13, 1981

Chinese Shenyang FT-6 two-seat version of the Soviet MiG-19 fighter, which entered service with the Pakistan Air Force last September, is a highly agile aircraft that is reducing student pilot checkout time sharply for the Shenyang F-6 fleet.

The FT-6 is being produced at Shenyang and Tianjin (Tientsin), where factories are geared to combined output of about 60 aircraft per month. It bears little resemblance to the MiG-19UTI trainer produced in limited quantities in the Soviet Union, where the MiG-19 is being replaced by the MiG-23.

To provide the second seat, the Shenyang engineering team increased the length of the forward section by 33 in. while retaining a single 30-mm. cannon beneath the nose for weapons training.

The FT-6 aircraft flown at this air base, with Wing Commander M. H. Dotani in the front cockpit, has an excess of specific power, which gives the trainer an impressive takeoff and climb performance. Pakistani F-6 pilots claim the fighter version can outclimb the Lockheed F-104 at medium altitudes.

During this flight, for example, the FT-6 was put into a sustained 6g turn at 20,000 ft. in which the aircraft accelerated from about 323 kt. to 453 kt. with use of afterburning in its Chinese-built Tumansky RD-9B-811 axial flow turbojet engines.

Dotani and other pilots said the FT-6 high performance is due to a thrust/weight ratio of more than 0.86:1. The two engines produce a total of 14,400 lb. thrust with afterburning.

Modifications in China have improved the view from both pilot seats and, since the fuel capacity was critical, two massive 30-mm. wing-root cannon have been removed and their mounting bays used for extra internal fuel.

This restores internal fuel to within 40 U.S. gal. of the single-seat fighter's 573 U.S. gal., about 2 min. less endurance. Even so, average safe sortie time in the FT-6 is only about 45 min. and after that, endurance is invariably critical.

Visibility from the slightly raised rear cockpit is better than that of the FT-5 trainer version of the MiG-17. The rear cockpit also has its own molded semicircular windshield for slipstream protection in an emergency. Chinese ejection seats, which feature face-blind actuation, are of limited performance.

Martin-Baker has studied a replacement seat along the lines of its Mk.10 zero-zero rocket boosted seat being retrofitted into the F-6 fighter. Small numbers involved in the FT-6 conversion, however, may make the program too expensive for the Pakistan air force.

Cockpit layout is similar to that of the FT-5 and has only basic avionics. The F-6 and FT-6, however, have fully powered ailerons and stabilators from independent hydraulic systems. Provision is made for manual reversion in the lateral control circuit. Controls include small interceptor spoilers that extend below the wing under-surface forward of downgoing ailerons to improve the rate of roll, particularly at supersonic speeds.

The stabilator circuit incorporates an electrical backup system for emergency operation, with Q-feel input to provide variable ratio gearing automatically between the stick and control surface according to speed and altitude.

This limits stabilator operation at high speeds and low altitudes and provides a natural change of effort and stick force per unit of g force, according to flight conditions. Manual reversion is demonstrated to students during training, but since control response is then very sluggish and heavy, exercises in this mode do not extend to landing.

The FT-6 also has pneumatically actuated brakes from a hand lever on the stick, and the bag-type brake units on the earlier Chinese aircraft have been replaced by disks, along with tubeless tires to improve ground control.

Pneumatics also are used for several other functions, including inflation of the hinged canopy pressure seals, gun charging and g-suit operation.

Gear and flaps are hydraulically operated with three pushbuttons to select zero, 15 deg. and 25 deg. extension of the slotted Fowler-type flaps. Half flap setting is cleared for use up to 431 kt. for higher maneuverability.

Afterburner nozzles are operated hydraulically but on takeoff burners normally are not turned on until the FT-6 is airborne, to avoid possible asymmetric effects resulting from uneven lightups. Maximum dry thrust of 4,740 lb. per engine is achieved at 100% rpm (11,150) but increases to the military power ratings of 5,730 lb. at the same rpm with tailpipe nozzles closed.

At this setting, takeoff acceleration was impressive on Mianwali's rough runway surface, with the rudder soon becoming fully effective for directional control at about 75 kt. Rotation was initiated at about 108 kt. and initially held to only 5-7 deg. to raise the nosewheel and avoid excessive angle of attack.

Further rotation at about 161 kt. resulted in unstick after a ground run of less than 3,000 ft. This could have been reduced to about 2,200 ft. at maximum gross weight of 22,000 lb. by earlier use of afterburning, but despite the close-set engines, the asymmetric safety speed is not achieved until the aircraft reaches 243 kt.

The FT-6 acceleration is such that after selection of afterburning the landing gear must be retracted quickly and locked in the up position before exceeding the safe limit speed of 269 kt.

It is possible to climb, using afterburners, at more than 30,000 fpm and to reach 56,000 ft. or more for high-attitude interception, but the required angle of climb at 377 kt. is uncomfortably steep to hold. The vertical speed indicator in the FT-6 has dual calibration from 0-20 meters/sec. and from 20-300 meters/sec. (3,937-59,055 fpm) and most of this range was explored during the flight with Dotani.

The flight did not extend to supersonic operations, although apart from a moderate nose tuck at around Mach 0.9, handling is considered unexceptionable up to a true Mach number of 1.45—limited by the thick wing and pitot intake—with little trim change or buffet in the transonic region.

Wave drag increases sharply from about Mach 0.9 to a first peak at Mach 1.18, then increases slightly to a second peak at the limiting Mach number. At the limit speed of 917 kt. at 35,000 ft., the maximum roll rate remains more than 50 deg./sec. due mainly to spoiler activation, although this rate more than trebles at subsonic speeds.

The FT-6 has a Sino-Soviet dual needle airspeed indicator, incorporating true airspeed up to 1,079 kt., but the aircraft is not equipped with a Mach meter.

At high subsonic speeds and medium altitudes, FT-6 handling was excellent, with light and responsive controls and wide buffet boundaries. Pakistan air force pilots are confident of outmaneuvering most other contemporary fighters in their F-6s through outstanding turn rates beyond, as this pilot experienced, the limit of their individual tolerances, even with g-suits.

This can be done without buffet or loss of airspeed up to the clean airframe limit of about +8g, but F-6 pilots would be ill-advised to slow down to meet lower-performance fighters on their own terms.

Despite the big wing fences, which help reduce spanwise flow and pitch-up at high angles of attack, lateral stability and control decrease markedly at lower airspeeds, where coarse use of aileron can flick one wing into a stall.

With experience, however, air force pilots said the F-6 can be flown down to very low airspeeds, using rudder for lateral control, although in the two-seat version students are not taken nearer than about 188 kt. to the clean stall of some 126 kt. because of the high sink rates likely to be encountered.

With full flap and gear extended, minimum flight speed is only about 6.4 kt. less, but in either case, the stall results after light airframe buffet and is accompanied by a pronounced wing drop. This obviously could lead to spin departures from turns at low speeds and high altitudes, but although intentional spins in the F-6 are prohibited, air force test pilots have identified from flight research three distinct modes to establish correct recovery procedures.

Whether in a normal or accelerated erect spin or an inverted spin, the aircraft has autorotational characteristics typical of swept-wing jet aircraft, with periodic snap oscillations and a tendency toward engine flameouts, particularly at high altitude.

In a typical erect spin, the aircraft stabilizes after three or four turns in a 60-70 deg. nose-down attitude and loses anything from 1,500 to 2,500 ft. per turn. In an inverted spin, the horizon apparently is not visible to the pilot, making determination of rotation very difficult except from the turn and slip indicator, and negative g forces may be sufficient to cause redout.

Standardized recovery procedures for an erect spin include firm application of neutral rudder and stabilator, followed by counter-spin aileron until rotation stops. Ailerons should then also be neutralized, and the aircraft recovered from the ensuing dive.

Spin avoidance

This method avoids the likelihood of entering an opposite or inverted spin if rudder is applied and not taken off soon enough, but the importance of the central white line on the FT-6 instrument panel as a neutral datum is emphasized by the warning that if more than 10 deg. of opposite aileron is applied, spin recovery may be prevented.

Inverted spin recovery involves determination of direction from the turn needle before firm application of full opposite rudder and full aft stick. The rudder pedals must be centralized as soon as rotation stops, before recovery from an inverted dive. Since about 7,000-10,000 ft. is required for safe recovery, immediate ejection is recommended in the event of an inadvertent spin below the latter altitude.

During aerobatics, it was noticeable that FT-6 maneuvering stick forces remained moderate throughout the speed range, with very little rudder required except to counter adverse aileron yaw at relatively low airspeeds. From a 485-kt. entry at about 12,000 ft. and 10,000 rpm, a 5½g loop topped out at 22,000 ft. and 243 kt. with virtually no buffet.

Engine handling was uncritical and apparently remains so even at very high altitudes. Relights also present few problems, and a brief exploration of asymmetric handling with one engine throttled back to flight idle showed little change of trim at normal operating speed. The F-6 fighter can complete a go-around from landing on one engine with care and some anticipation.

Trim is achieved by individual electric actuators adjacent to the throttles via aileron and rudder tabs, or through the stick datum for the stabilator circuit. The FT-6 tends to run short of longitudinal trim at speeds above about 458 kt., but the nosedown pitch between Mach 0.98 and 1.2 can be held easily on the stick.

Trim changes are otherwise small, with little effect from extension of the big three-segment air brakes flanking and below the center fuselage. These can be left open for landing, for which the first 15 deg. of flap may also be lowered below 215 kt., with a slight nose-up pitch.

A normal fighter break from about 323 kt. overhead proved effective in bleeding off speed for the landing pattern, in which engine rpm was maintained well above the required 7,500 to insure rapid acceleration. Power was used to continue a fairly shallow descent of just under 1,000 ft./min., reducing to 188-205 kt. on final with full 25 deg. of flap at around 8,000 rpm.

Induced drag of the 55-deg. swept wing is high in the landing configuration so the change of attitude in the 161-kt. flare-out at 10-15 ft. was fairly small, and throttle reduction to idle was delayed until just before touchdown at 145 kt. to avoid excessive sink rates. Slower approach speeds are not recommended because of stability and wing drop problems, high sink rates and the danger of touching the ground with the rear fuselage at higher angles of attack.

Stabilator effectiveness is such that the nosewheel may be held off the ground for aerodynamic braking, but the normal technique is to lower it below about 129 kt. This is also the limiting speed for deployment of the tail braking parachute used for every F-6 and FT-6 landing, and which in earlier aircraft with under-fuselage stowage resulted in a strong nosedown pitching moment.

All three wheels, therefore, had to be on the ground before chute deployment, but this is unnecessary with later F-6s and all FT-6s. These have their pneumatically released tail parachutes stowed in a streamlined fairing above the fuselage at the base of the rudder, allowing earlier deployment. This also alters the distinctive rear fuselage contours of the MiG-19, following removal of the characteristic "pen-knib" fairings above and below the twin tailpipes, and resulting in increased yawing forces, especially in crosswinds.

It still is advisable to get all three wheels on the ground as soon as possible, however, since the MiG-19 is one of the few aircraft with a separate nosewheel brake. This is operated independently through its own hydraulic circuit and undoubtedly helps reduce the landing roll, although to what degree is uncertain.

"Saving Grace"

As an air-superiority fighter, with the addition of U.S. air-to-air missiles, the F-6 is still regarded by the Pakistan air force as equal to most of its opposition, and its supply by China has been described by Air Chief Marshal Muhammad Anwar Shamim, chief of staff, as "a saving grace" for Pakistan. Its maintenance requires a great number of man-hours, but its spares are inexpensive, and it is one of the least costly combat aircraft to operate in its class.

Workmanship and finish of the Chinese-built F-6s and FT-6s are impressive, having improved considerably over the past few years, although the service lives of some components are short by Western standards. This is particularly true of its turbojet engines, which have a TBO of only 100 hr., and the Pakistan air force would like to reengine its 150 or so F-6s with longer-life and lower-consumption turbofans. Something like the Rolls-Royce/Turbomeca Adour, it is believed, would make the F-6 an outstanding fighter, but reengining would take too long, cost too much and require a more efficient intake to be practical.

Chapter 3

Cargo/transport
&
special purpose

C-17 should fulfill
USAF airlift mission

David M. North/Edwards AFB, Calif.
May 10, 1993

The Air Force sees the Globemaster 3 as the one aircraft capable of providing needed future airlift in a changing and expanding mission and replacing its aging current airlift fleet. At the some time, the ineptitude of McDonnell Douglas in bringing the aircraft to this point in its development, plus a declining defense budget, has placed the program under intense political scrutiny. The more than $1.3 billion in cost overruns, the two-year delay in aircraft deliveries, the alleged program mismanagement and the failed wing structure have made the C-17 an easy target for its opponents. If the C-17 were cancelled today, the Air Force would have to reinvent the aircraft at a greater cost and delay. McDonnell Douglas still has a credibility problem with the program, but it has lowered the production costs and time, and is addressing the technical problems. This is not to say that the C-17 meets all of its technical goals and performance parameters at this time, but as in any concurrent program the fixes are still being accomplished. This pilot report will not recount the C-17's past problems or achievements: Aviation Week & Space Technology *has done that over the past 10 years (AW&ST Mar 22, p. 28; Mar. 15, p. 30). Instead, the report will detail the experiences of the first pilot to fly the aircraft other than a McDonnell Douglas or Air Force test pilot or a four-star Air Force general.*

The McDonnell Douglas C-17 should meet the U.S. Air Force's requirement for an aircraft to perform the combined long-range strategic and short-range tactical airlift roles.

In the early 1980s, the Air Force started to look for an aircraft that could carry outsize equipment and land on short, unpaved runways. The aircraft had to be self-sufficient on the ground, use minimal parking space and be able to airdrop troops and equipment. The new transport also had to have ample range with a good size payload, that combined with inflight refueling could reach almost any point in the world.

The current airlift aircraft in the U.S. Air Force inventory—the Lockheed C-5, the C-141 and the C-130—all perform selected elements of the requirements, but not all combined as the C-17 is in-

McDonnell Douglas C-17 is powered by four Pratt & Whitney F117-PW-100 engines. The U.S. Air Force plans to buy 120 of the airlift aircraft.

tended. The airlift fleet also is becoming older. The average age of the C-141B fleet is 28 years, the C-5As average 22 years and the C-5B average six years.

Viewed from the cockpit, the McDonnell Douglas C-17 meets the Air Force's overall requirements that were defined in the 1980s. That is not to say that the P-4 Globemaster 3, flown by this *Aviation Week & Space Technology* pilot on Apr. 27, meets all those requirements at this time. There were limitations placed on the flight because of technical problems and because the flight test program has not opened up all of the C-17's performance envelope. As with any concurrent flight test program, the manufacturer and the customer are working to solve these problems.

One other journalist pilot and I were allowed to fly the C-17 in the middle of the test program. I had put in a request almost a year ago. The Air Force at first notified me that the flight would take place in January. Then because of software changes to the digital flight control software, the flight was delayed to June. I was notified early in April that the flight would take place in late April.

Although the reason for the schedule shift was not given, it was apparent that the C-17 was coming under extreme political pressure because of cost and schedule overruns, alleged mismanagement and technical difficulties. A positive pilot report from an outside observer would help counter the perception that the C-17 was a technical disaster and could not do the Air Force mission. When I explained to a senior Air Force officer that there was no guarantee of a positive pilot report, the reply was "we will take that chance."

In taking that chance, in my view, the Air Force was right. The C-17 delivered to the Charleston AFB this year will be able to do the

Air Force mission and do it well. Because the aircraft is still in flight test, there are still potential technical and performance problems. For example, the C-17 has only been tested into the stall regime with a forward center of gravity. It will undergo flights with mid and aft cg. conditions later this year.

The flight from the test facility here was in the No. 4 production C-17 and it was the aircraft's 23rd flight. McDonnell Douglas C-17 chief test pilot Chuck Walls was the designated pilot and was to occupy the right seat while I took the left seat. Air Force Lt. Col. Kermit Rufsvold was the safety officer and observer. Robert Ainsworth was the flight test engineer from the manufacturer. Bill Yeary and Ted Venturini were the two loadmasters.

Walls performed the outside preflight and detailed some of the C-17's features. The auxiliary power unit is in the forward section of the right landing gear pod. This installation makes the right pod longer than the left, but it did not appear to create an imbalance in flight. Walls said there was no need to clear the APU for use in flight.

C-17 cockpit is equipped with four multifunction displays, three mission computers and GEC Avionics HUDs. The HUD was used throughout the evaluation for primary flight reference.

The nose-wheel gear system is being modified to include a second actuator for retraction in parallel with the current single actuator. The test pilots are finding that raising the nose gear at a speed higher than 170 kt. sometimes results in an unsafe gear indication.

This was a problem during the C-17's first flight in September, 1991. The single actuator does not have the power to fully raise the gear that retracts forward. The attachment points were installed, but the actuator itself will be added later.

The C-17's large size was highlighted by a C-141 parked nearby. The C-17's length and wing span are within 10 ft. of the C-141; however, the fuselage diameter is within a foot of the C-5.

Rufsvold had started the aircraft's APU prior to our boarding the aircraft and had most of the systems on line. There was a slight delay in starting the engines while the flight crew attempted to get all of the displays and mission computers to agree. There are three identical mission computers in the C-17, and Walls said that in the flight test program, there has been some difficulty getting them into synchronous operation.

This portion of the avionics suite was designed early in the program, and operations have shown that it might have been better to distribute the computer's functions by using different system architecture.

Once the displays were functional and the avionics systems were in agreement, the Pratt & Whitney F117 engines were started. The automatic sequence involved pushing a button on the overhead panel and monitoring the engine parameters with the throttles in idle. It took close to 2 min. to start the large fan engines. Walls said two engines could be started simultaneously to reduce the overall time.

An automatic sequence of internal checking of the spoiler, slat and flap controls required 8 min. while not touching any of the flight controls. In a later software update, the sequence time will be reduced and the manual operation of switches in the overhead will be replaced by a single test button.

The gross weight of the aircraft on the ramp was 431,200 lb., some 153,800 lb. less than the 585,000-lb. maximum allowable weight. The basic empty weight was 274,400 lb., and there was 119,900 lb. of fuel on board. The weight of the eight-person crew was 1,600 lb., and ballast in the front of the cargo compartment was an additional 35,300 lb.

Fuel flow at idle power was 3,900 lb./hr. I added power to start the movement from the ramp and then used the nose-wheel handle to the left of the pilot's seat to maneuver. A digital readout on the primary flight display allowed taxi-speed monitoring. The brakes were effective and provided smooth deceleration at the full range of taxi speeds.

Prior to reaching the runway, Walls had me slow to a 5-kt. taxi speed and perform a 360-deg. turn to assess the C-17's ground handling capability. The aircraft turned almost on its left gear using only the nose-gear steering and not brakes. The ability of the aircraft to make tight turns and to back up using engine thrust reversers is seen as a positive factor when parking a number of aircraft on small ramps. The C-17 has been backed up a 2-deg. grade during tests.

"This is not just a demonstration feature," Rufsvold said. "Reverse will be used operationally all the time." We were not able to try the reverse thrust because during flight test it was found that heat from the engine exhaust caused dimpling to the slats. The slats near the engines will be constructed of titanium, rather than the current aluminum.

The V_r rotation speed had been calculated to be 122 kt. and the V_2 takeoff safety speed was 139 kt. The minimum retract speed for the half-flap setting was 159 kt. and the slats were to be raised at 200 kt. Ainsworth calculated the balanced field length to be 4,800 ft. The wind was from 080 deg. at 4 kt. I lined the C-17 up on Runway 22 at Edwards and advanced the throttles to a 1.2 engine pressure ratio (EPR) setting. The throttles were advanced to the takeoff limit as calculated by the digital electronic engine control and was in the 1.5 range. The electronic engine control does not allow the setting to go past the limits. The temperature on the morning flight was 67F at the 2,310-ft. field altitude.

Acceleration was brisk and nosewheel steering through the rudder pedals was used on initial roll prior to rotating to a 10-deg. attitude after a ground roll of 4,000 ft. The landing gear was raised prior to 170 kt., and the flaps and slats retracted at the appropriate speeds.

The four-channel digital flight control system was immediately noticeable during the initial climb. Once the climb was established, I was able to take my hands off the stick; roll and pitch stayed constant. As a former Navy attack pilot, and having flown with a control stick in various aircraft, I preferred the stick in the C-17 to the traditional yoke. The installation allows an uninterrupted view of the primary flight display and is more precise in controlling the aircraft. The stick moves fully aft and forward in pitch control, and pivots on a point below the grip in roll control.

Within 4 min., the aircraft was climbing through 10,000 ft. at 3,500 fpm at 238 kt. Fuel flow was 36,600 lb./hr., and the EPR setting was 1.35. Passing through 15,000 ft., the C-17 was accelerating to 300 kt., with a fuel flow of 32,600 lb./hr., and it had taken 6.6 min. to reach this altitude.

Ten minutes after takeoff, the aircraft passed through 20,000 ft. at a 1,900-fpm climb and burning 28,800 lb./hr. Another 4.1 min. was required to reach 25,000 ft. where the rate of climb was 1,060 fpm and the speed 309 kt. The rate of climb decreased from this point to the cruising altitude of 33,000 ft. because of turns required to stay within the operating area.

The altitude of 30,000 ft. was reached in 21.1 min. from takeoff, and the rate of climb was 1,100 fpm. Fuel flow was 22,400 lb./hr. and the speed was Mach 0.78. It took 26.7 min. from takeoff to reach 33,000 ft. at the established Mach 0.78 speed. Total fuel used was 13,420 lb. from takeoff.

During the climb to the cruising area, I used the GEC head-up display (HUD) as the primary reference for headings, speeds and altitude. Unlike with some HUD displays, I found the digital vertical speed readout easy to assimilate into the scan pattern. The horizontal

McDonnell Douglas and the Air Force flew six C-17s in the aircraft flight test program.

plan of the operating area was shown on our multifunction displays. The primary flight display, engine normal readouts and configuration layout were shown alternately on the other two displays in the center of the instrument panel.

At a cruise speed of Mach 0.775, fuel flow was 17,000 lb./hr. The flight computer said that at the current gross weight of close to 417,000 lb., the aircraft could cruise at 34,000 ft. During operational flights, the crews would step-climb the aircraft to higher altitudes, as gross weight decreased with fuel burn. While at cruise Mach, and using less than a third of stick movement for roll control, the rate was positive and brisk. The pitch movement corresponding to control input was immediate but with a slower rate, as would be expected in a large cargo aircraft. This was to be true throughout the flight, and I found aircraft response to be in complete harmony with flight control movement.

One of the primary performance issues surrounding the C-17 is the range and payload figures. McDonnell Douglas promised higher numbers than initially required by the Air Force, and the agreed contract specifications call for a maximum range of 2,400 naut. mi. with a payload of 160,000 lb. with established reserves. The shortfall in payload when meeting the specified range is close to 9,800 lb., while the range is almost 220 naut. mi. short when carrying the 160,000-lb. payload.

McDonnell Douglas has instituted a weight reduction program to eliminate 1,435 lb. The maximum takeoff gross weight is being upped by 5,000 lb., and the company is trying to identify aerodynamic areas to decrease drag by 1%.

Pratt & Whitney and the contractor agree that the total specific fuel consumption of the F117 engines is 2.5% high. With identified upgrades to the PW2040 commercial engine passed on to the F117, the Air Force will gain a 0.6% decrease in specific fuel consumption.

While additional improvements are possible, the Air Force would have to break away from the commercial engine specifications, and suffer higher spare parts costs and possible reliability and maintainability cost increases. At this time, the service wants to stay compatible with the commercial engine, which has accumulated more than 4.5 million flight hr. in the Boeing 757.

"If you ask an Air Force operator whether he wants to give up the thrust reversers, or some cargo kits to gain 200 naut. mi. to reach an arbitrary range figure, he will tell you to forget it," one senior Air Force officer said. "The current fuel reserve requirements are too high and not realistic for the C-17."

The Pratt & Whitney F117 engines in the C-17 flight test aircraft had gone without an unscheduled removal until a month ago. One engine had to be removed when a carbon seal overheated and failed. The failure came when the aircraft was undergoing a series of negative-g maneuvers. The engine manufacturer is testing the carbon seal, but also is trying to determine if the maneuvers have an operational use.

Prior to my C-17 flight, I had spent 1.5 hr. with Walls in the flight hardware simulator in Long Beach, where he demonstrated many of the automatic functions of the mission computers and autopilot. The navigation system is coordinated by the mission computers using conventional navigation radios, TACAN, four inertial reference units, weather radar and a global positioning system. The autopilot is able to fly coupled mission computer-generated approaches with autothrottle engaged. During the simulator flight, all of the automatic modes appeared to function well. Walls said the vertical navigation profile had not been perfected yet, but that it was under development.

I then retarded the throttles and lowered the C-17's nose to achieve a maximum Mach of 0.825 at 30,000 ft. While there was an aural warning of overspeed, there was no buffet in the aircraft. As we descended through 27,000 ft., the speedbrakes were deployed with an initial airframe buffet, which quickly subsided. At 18,000 ft., the rate of descent was 8,000 fpm and the speed 340 kt.

The leading edge slats were deployed and a 250-kt. speed was established at 15,000 ft. I had requested to look at the performance of the aircraft in a normal inflight refueling situation, although there was no tanker present. Power response from the four F117 engines was positive at this altitude. Engine power response from idle to full power at this altitude was approximately 6 sec.

I found when not monitoring engine power, I often overcorrected for airspeed changes. This was partially due to the long throw of the throttle handles. Roll, pitch and yaw control was effective and by lining up the aircraft's nose on a distant point, I was able to judge the movement of the C-17 as if flying station on a tanker. Walls said the current flight control software had improved aircraft performance during inflight refueling, and most of the test pilots found the maneuver relatively easy to execute.

Stalls would normally have been performed at approximately 15,000 ft. Because of test restrictions, we were unable to do in-flight stalls, but they had been done in the simulator. The simulator was equipped with an attitude limiting system (ALS), which prevents further aft stick movement when an appropriate stall speed is reached. The system is being evaluated in one of the test C-17s.

J.D. (Doug) Burns is a McDonnell Douglas test pilot who has flown much of the C-17 high-angle-of-attack testing. He said that, in general, with a forward center of gravity, the C-17 is controllable into the stall with no tendency to pitch up, even when at a maximum of 35-deg. AOA in the clean configuration. At the lower thrust levels, Burns said, there is less buffet into the stall in the clean configuration, while with higher thrust there is less buffet with flaps extended. The ALS will have its "soft" limit at the stick shaker speed of 1.15 to 1.05 of V_{stall} (low stall). The "hard" limit of the ALS will be at the maximum coefficient of lift.

The only surprise so far in the stall work, Burns said, was that with flaps extended the stall speeds are 2-4 kt. higher than predicted. All of the stall testing has been with forward center of gravity, and with no tendency for a deep stall. "There is potential for a deep stall in the mid and aft cg. regimes, but I do not think it will happen. However, this is why we installed the ALS."

The C-17 was then flown to 7,500 ft. and slowed to 130 kt. with the flaps extended to the ¾ position and the index at 96%. At this point, the loadmaster attempted to open the right troop door but found it difficult to do with the existing door mechanisms. Once the doors were opened, the wind deflectors were deployed and the cargo ramp was open to simulate air drops. There was no change in the aircraft's flight characteristics while holding 130 kt.

Later in the flight, I went down to the cargo hold during a similar open-door maneuver at 130 kt. With the crew chief, I was able to stand in the middle of the cargo ramp with little airflow. The only location where there was some airflow was near the side of the cargo compartment between the troop doors and the open cargo ramp.

I then turned toward Edwards to perform landings. Again, because of the higher than anticipated temperatures of the exhaust gases through the blown flaps, there was a restriction on the use of full flaps for landing. The titanium flaps are due to be installed later this year for flight test. The maximum flap setting was 30 deg. with an index of 96%.

At Walls' suggestion, I established a 5-deg. nose-up attitude on the downwind to Runway 22. The pitch hold mode for the digital flight control was selected by depressing a button on the stick and was verified on the HUD. The landing gear and partial flaps were deployed while downwind, with final flaps selected at the 90-deg. position.

Flying the C-17 on the backside of the power curve was much like the technique used for carrier landings. As in a carrier approach, the aircraft is flown to touchdown, without flaring to decrease the landing impact. Walls estimated that ground effect lowered the landing vertical speed by about 100-200 fpm.

Aft loadmaster's station has computer controls and backup controls. The hydraulic valves can be manually operated through the cutout at lower right.

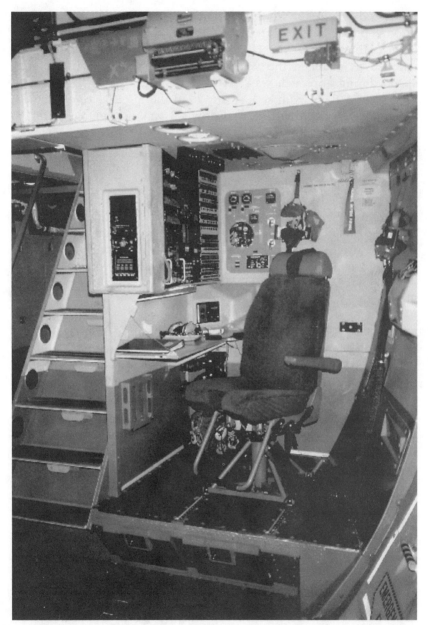

Forward loadmaster's station has the cargo control panel facing the loadmaster, and any oxygen control board to his right. Red guarded switches are hard-wired electrical backups to the computer. The panel has a TV monitor to verify proper opening of airdrop parachutes.

Speed was set by the pitch, and altitude control was achieved through power changes. From the 1,000-ft. point on final, I placed the flight path vector in the HUD on the end of the runway and attempted to maintain the vector on the same spot. The reference speed during approach was 131 kt. for the 2.5-deg. glideslope approach. The flight path was steady until the last 100 ft. of altitude, when the velocity vector started down. I was not quick enough to catch the vector with power, so the landing was firm. Gross weight of the aircraft was 383,000 lb. on the first landing.

The second approach was a tight lefthand pattern to the same runway. Visibility from the cockpit is excellent through the main windows and aided by an eyebrow window and a lower large ground observation window. This time I established the 5-deg. attitude and kept the HUD vector on the 2.5-deg. glideslope reference to touchdown by use of power. The touchdown was much smoother.

Walls retracted power on the No. 4 engine on the downwind, and it would have been undetectable had it not been for a thrust-loss light illuminating near the HUD.

In flying the C-17, as in many modern tactical aircraft, the pilot becomes almost completely dependent on the HUD for primary flight information. The responses made in power settings are dictated by HUD information without reference to the engine instrument displays. The digital flight control system compensated for yaw with the engine out.

The third landing was much the same as the second, even with the engine out. On landing, the throttles were retarded to idle, and the four engines were put into idle reverse using the handles mounted forward on these throttles, which I found easier to manipulate than using the throttles themselves.

Walls later demonstrated a short-field landing without the use of full flaps and reverse thrust. The approach speed was 125 kt. at the aircraft's 350,000-lb. gross weight. Touchdown was at 120 kt., and with full braking, Walls stopped the aircraft in less than 2,800 ft. He said that with use of full blown flaps, the speeds would have been 15 kt. lower.

Total flight time was 2.5 hr., including three landings and much of the time spent at lower altitudes demonstrating the aircraft's air drop capabilities. The fuel used for taxi was 3,000 lb., and the total fuel used from engine start to stopping on the taxiway was 55,300 lb.

Aircraft P-4 had a number of nuisance faults during flight, including stall warnings when not in stall conditions. There was a failure of the heading select function of the autopilot during flight. However, when these faults are measured against the complexity of the aircraft, they seem minor.

More of concern, however, are the technical problems that limited what we were able to accomplish during the evaluation flight. The lower nose-gear retraction speed because of the actuator, and the inability to use reverse thrust and full flaps because McDonnell Douglas did not correctly estimate the effect of the engine exhaust on the slats and flaps, are key examples of technical difficulties. The failed-wing-related flight restrictions also are delaying the development program.

If McDonnell Douglas has accurately established the fixes for these problems—and moves quickly to fix any further problems identified during flight test—then the Air Force will receive a good aircraft to fit its mission.

Versatility and automation
are key to C-17 cargo operations
Michael A. Dornheim/Edwards AFB, Calif.

The Air Force/Douglas C-17 will be more prepared to perform a variety of missions than existing transports as a result of extensive input from cargo loadmasters in the design process.

Items that are optional kits on other aircraft are standard equipment on the C-17. This results in a heavier aircraft with less payload-range capability, but Air Force users are willing to trade that for a versatility that is always on board.

The C-17 cargo system is more automated than its predecessors, allowing one loadmaster to do an airdrop that would take three people in a Lockheed C-5. The system also has two levels of backup in case the automation goes awry due to computer problems or combat damage.

The backup was demonstrated during this *Aviation Week & Space Technology* editor's tour of aircraft P-4's cargo hold. Theodore R. Venturini, Douglas C-17 chief project loadmaster, had used the cargo system computer to command the aft ramp to lower on the ground but the ramp latch locks would only make clicking sounds. Several cycles of pressing buttons on the computer diagnostic panel implicated a position sensor that would not yield to electronic persuasion.

Msgt. William Yeary, the Air Force loadmaster, then inserted a lever to manually unlock the latches. Venturini then opened the hydraulic manifold panel and manually operated the valves to lower the ramp. The direct approach succeeded over the balky electronics.

Venturini has had long involvement with the C-17 design. In the 1970s he was the Air Force project loadmaster for the YC-14 and YC-15 transport prototype programs, from which the C-17 draws its design heritage. These programs assembled military cargo operations experience into their cargo systems design. Though not implemented on the test aircraft, these features were built into fuselage mockups. Venturini went to Douglas in 1979 to take part in C-17 design.

Key to the C-17's flexibility is that it carries a lot of equipment. The empty weight is thousands of pounds higher than it would be if the C-17 were fitted to C-5 standards, but Venturini will not give any of it up despite C-17 payload-range shortfalls. "I don't give an inch in weight," he said. "If you do, you will destroy it slowly but surely. Versatility is more important than payload in the field."

For example, "C-141s from Norton AFB made an operational readiness inspection flight to Germany," Venturini said. "Then they said 'let's do an airdrop'—but they didn't have the airdrop kit. They had to fly aircraft in from McGuire AFB to bring in the kits."

Standard C-17 equipment includes retractable floor rails for carrying a double width of 88-in.-wide logistics pallets, a single width of 108-in. air delivery system (ADS) pallets, or a double-width of 48-in. Army container delivery system parachuted pallets. These cargo systems can be mixed, and reconfiguration of the built-in rails and rollers is quick—the floor can go from flat to rollers in 27 min. Aerodynamic fences and other provisions for airdropping supplies and troops are built-in. On the C-5, the ADS rails and aerodynamic fences are separate kits weighing over 7,000 lb.

Struts hydraulically extend from beneath the ramp hinge for stabilization during loading, and dual fans blow fresh air to clear the compartment of vehicle fumes. Stanchions for 12 litter patients are neatly stored on the wall to always be ready for medical evacuation, along with oxygen and electrical outlets for 48 patients. There is also oxygen for 54 high-altitude paratroops.

Automation is symbolized by the loadmaster's station at the front of the cargo bay, which can control all aspects of airdrops, and combat off-loads while taxiing. Electrical actuators change the cargo locks into spring detents for airdrops. A remote television camera lets the loadmaster check the proper deployment of parachutes out the back while releasing an airdrop. The C-5's manual airdrop operation takes three people moving through the cabin, who can be tossed about by aircraft maneuvers.

There is a smaller stand-up control station by the back of the aircraft, as well as several small panels along the fuselage for local control of cargo locks. As a backup to the cargo computer, guarded switches give direct electrical commands to the hydraulic valves. The next backup is manual activation of the valves, as Venturini and Yeary demonstrated. The cargo system is powered by the aircraft's No. 2 hydraulic system. If it is damaged, there is an isolated sump with enough fluid to close the cargo ramp.

C-17 SPECIFICATIONS

POWERPLANT
Four Pratt & Whitney PW2040 (military F117-PW-100) engines with 41,700 lb. of thrust each.

WEIGHTS
Maximum gross weight	585,000 lb. (265,590 kg.)
Maximum payload	172,200 lb. (78,109 kg.)
Approx. empty weight, No. 4	274,400 lb. (124,578 kg.)
Fuel capacity	176,200 lb. (79,923 kg.)

DIMENSIONS
Length	174 ft. (53.04 meters)
Wingspan	171.2 ft. (52.2 meters)
Height at tail	55.1 ft. (16.79 meters)
Wheel to wheel (outside)	33.7 ft. (10.3 meters)
Fuselage diameter	22.5 ft. (6.86 meters)
Cargo floor length	68.2 ft. (20.8 meters)
Ramp length	19.8 ft. (6.04 meters)
Loadable width	18 ft. (5.48 meters)
Cargo floor height	12.3 ft. (3.75 meters)
Wing area	3,800 sq. ft. (353 sq. meters)
Aspect ratio	7.165

PERFORMANCE
Range/160,000 lb. payload	2,400 naut. mi.
Cruise speed	Mach 0.77 at 28,000 ft.
Ferry range	4,600 naut. mi.
Service ceiling	45,000 ft.
Takeoff length/MGW	7,600 ft. (2,318 meters)
Landing length/max payload	3,000 ft. (915 meters)

VISTA primed for research, training

David M. North/Wright Patterson, AFB
March 6, 1995

U.S. Air Force and Navy test pilots will be among the first to evaluate the variable stability aircraft's aeronautical research and training capabilities.

U.S. Air Force and Calspan's NF-16D, the newest variable stability aircraft, is now ready to offer operators a vehicle to perform aeronautical research and the training of new test pilots.

After a start-and-stop beginning in the late 1980s and early 1990s, the modified Lockheed F-16 variable stability in-flight simulator test aircraft (VISTA) is now available for long term flight test commitments. The Air Force's Flight Dynamics Directorate here, manager of VISTA, also has plans to incorporate new features into VISTA, which will increase its capabilities as a research aircraft.

The plans for the NF-16D include flight control development and evaluation flights for the Lockheed F-22 and the Indian Light Combat Aircraft (LCA). During this summer, the VISTA will be flown with the F-22 flight system model to evaluate the calibration. A more detailed F-22 flight test program is scheduled for early 1996. Development of the flight control system for the Indian LCA is expected to start in late 1995. The Indian air force is expected to send two pilots to fly with Calspan in the VISTA. Initial development flights for the LCA will be flown this spring in the Lockheed NT-33. The NT-33, in operation since the mid-1950s, is being replaced by the NF-16D.

More immediately, VISTA is scheduled to visit the Air Force Test Pilot School (TPS) at Edwards AFB, Calif., this month. At that time, instructors from the test pilot school will be flown in the NF-16D to demonstrate the capabilities of the aircraft. Later in the year, TPS students will receive two flights in the aircraft to evaluate a wide range of flight handling qualities.

Following the visit to Edwards AFB, the VISTA will be flown to the Naval Test Pilot School at Patuxent River, Md. Each student at the Navy school, including fixed wing and rotary pilots, will receive one flight in the variable stability aircraft.

The flight that this *Aviation Week & Space Technology* pilot flew from here closely resembled the syllabus flight that will be flown by the Navy students, Jeffrey H. Peer, chief test pilot for Calspan's Flight Research Dept., said.

To expand the aerodynamic research capabilities of the VISTA, the first project to be undertaken by the Flight Dynamics Directorate is to return a multi-axis thrust vectoring engine to the F-16 airframe. The VISTA is the same aircraft that had been used by the joint Air Force/Lockheed/General Electric team to conduct the MATV program (AW&ST May 2, 1994, p. 46).

The air force recently selected the Pratt & Whitney PYBBN configuration of the F100-229 engine to power the NF-16D over the General Electric candidate. The General Electric F110-GE110-100 engine with multi-axis thrust vectoring was the one that was flown in the MATV program.

At this time, the Air Force expects to install the Pratt & Whitney multi-axis thrust vectoring engine in the NF-16D in the last half of 1996. The rest of 1996 will be devoted to computer software development for the aircraft. A six-month flight test program is to be conducted in 1997, and the Air Force and Calspan will have the aircraft ready for customers in 1998. Unlike the MATV program, which was restricted to the investigation of high angle of attack flight above 20,000 ft., the refurbished VISTA will be used to explore the entire flight envelope, including takeoffs and landings.

While the aircraft is being modified with the vectored thrust engine, the VISTA will be equipped with a programmable helmet mounted display in the front seat and a helmet mounted display in the rear seat.

USAF and Calspan's VISTA has removable hard drive disk that contains 200 preprogrammed flight control configurations.

New multi-function displays will be put in both cockpits, and a head-up display will be added to the rear seat. This configuration will allow a customer to fly almost any cockpit display in the aircraft that can replicate the performance of most current and new aircraft.

The Air Force selected the F-16 airframe for the variable stability role over a number of other aircraft, including the McDonnell Douglas F/A-l8, Northrop F-20 and the British Aerospace Hawk. A contract was awarded to Calspan in 1988 to install a center stick and integrate the computers needed to perform variable stability flights. The Block 30 F-16D, No. 86048, was supplied by the Air Force's Tactical Air Command. The VISTA F-16 flew five times in 1992, with the final flight the acceptance flight by the Air Force. The aircraft was then placed in flyable storage because of a lack of funding.

The variable stability computers were removed from VISTA for the MATV program, which began flight tests in July, 1993, and put back in when the MATV program was concluded in March, 1994.

Three Rolm Hawk computers, mounted in the aircraft's large dorsal fin, provide the variable stability functions and make VISTA an in-flight simulator. The computer system monitors the pilot's inputs and then moves the aircraft's control surfaces to produce the required mo-

Primary flight instruments had to be removed from the VISTA front cockpit for the installation of the Calspan center stick.

tions. The controls to access the computer to change flight characteristics and engage the front seat controls are mounted in the back seat.
The variable stability system (VSS) commands symmetric and asymmetric horizontal tail movement, symmetric and asymmetric flaperon
movement, rudder and throttle control. The only surfaces not controlled by the VSS are the leading edge flaps and speedbrakes.

Other modifications included in the basic VISTA configuration include heavy weight landing gear and a larger capacity hydraulic pump
and lines to accommodate the increased surface motions needed to
simulate other aircraft.

My opportunity to fly the NF-16D came two days after the Air
Force's Wright Laboratory accepted the VISTA aircraft from Lockheed.
The flight was flown from the Air Force facility here with Maj. Robert
A. Wilson, USAF, a test pilot from Edwards AFB assigned to the VISTA
project.

The preflight briefings also were provided by Calspan's Jeffrey
Peer. Peer and other Calspan pilots will fly the VISTA with company
pilots evaluating aircraft characteristics and with the students at the
service test pilot schools.

Fortunately, Wilson and I were able to strap into the aircraft while
it was still in the hangar on the cold Dayton day. I took the front seat
while Wilson occupied the rear seat, which is the command seat in
the VISTA. Peer and Wilson had explained to me many of the symbols that would appear in the multi-function display while evaluating
the variable stability system.

The aircraft was started once we were outside the hangar, and
Wilson performed the ground check of the VSS. Wilson then taxied
the NF-16D to the active runway and performed the takeoff. Once airborne, Walker turned over the controls to me. Taking control of the
aircraft from the front seat involves pushing an F-16 convenience button and then matching the front throttle to the rear throttle until there
is an indication in the front seat. Because the front throttle is not connected mechanically to the fuel shutoff valve, the engine cannot be
shut down from the front seat.

The VISTA F-16 is, if not the only one, one of a very few F-16s
equipped with a center control stick, plus the side-stick controller.
The center stick's functions are completely programmable from the
rear seat. The side-stick controller functions like a traditional F-16 in
the NF-16D. Later this year, Calspan will modify the front seat's side
stick with a variable feel system.

The installation of the center stick in the NF-16D forced the removal of some of the primary flight and navigational instruments. I

found that I used the aircraft's head-up display for flying and then for the instrument landing approach at Wright-Patterson AFB.

There are a number of safety trips that have been incorporated into the VSS for safety reasons. For example, aircraft control will revert to the rear seat when 7.33 positive Gs are pulled, when the speed exceeds 440 kt. and the angle of attack exceeds 16 units. A number of the other VSS maximum parameters were exceeded during the flight and control reverted to Wilson.

The first part of the flight evaluation involved flying the NF-16D as what Wilson described as a "reasonable handling" aircraft using the center stick. Wilson ascribed the handling characteristics to closely matching those of the Northrop T-38. I performed some sharp turn reversals, aileron rolls and a split-S maneuver. A loop during which I exceeded 440 kt. reverted the aircraft to Wilson. The multifunction display indicated that the overspeed safety trip had occurred.

The next mode flown was in symmetric flaps, which provided direct lift up and down. The flap movement of 2 deg. provided what I best could describe as "wobbly flight" in all axis.

An F-16 without flight control system augmentation was the next mode introduced by Wilson. The F-16's response to center stick input was sloppy at first, and then the input appeared to have a more positive response. Performing an aileron roll and a four-point roll, I found that the nose wandered and there was a tendency to over-compensate in fixing the wing position. Rudder input in this mode gave way to the nose describing a wide arc and a bobbing sensation.

Three Rolm Hawk computers control the movement of the horizontal tail and flaperons, plus the rudder and throttle, through the VSS.

Managing Editor David M. North flew VISTA with Maj. Robert A. Wilson from Wright-Patterson AFB.

"This is a flyable aircraft, but not one with good flight characteristics," Wilson said.

Wilson next put an adverse yaw mode into the flight control system. Quickly performing 60-deg. banked turns to both sides caused the aircraft's nose to initially slice in the opposite direction of the bank. This tendency could have been reduced by the input of coordinated rudder during the banked turn, a proclivity not normally found among F-16 or single-engine jet pilots.

Countering the F-16 without augmentation which was flown earlier, Wilson next put an overaugmentation configuration into the flight control system. "In theory this mode might look good to the design engineers, but is not acceptable to the pilot," Wilson said. I found that small control inputs from the center stick resulted in apparently larger control surface deflections, with an immediate sluggishness in aircraft response followed by over-response.

The overaugmentation was then tied to the side-stick controller and the lifting body capability of the NF-16D to put the aircraft into a coupled roll mode. A slight banked turn to the right that wanted to continue and the resultant opposite stick force quickly had me in a pilot induced oscillation. I was saved from further control problems by the safety feature kicking in and restoring normal flight.

A divergent spiral mode in the F-16 caused the fighter to try and continue to increase bank angle after control input was neutralized. This would be a difficult configuration to fly in weather conditions.

Wilson then incorporated a 200 millesec. delay in aileron response into the flight control system. While attempting to do 30-deg. bank angle captures, even with the small delay, I found it easy to get behind the aircraft, and then out of phase with the recovery until the safety system released me from a pilot induced oscillation situation. While aggressively trying to capture 30-deg. bank turns with a rate limit on the ailerons, I found myself flying into the same situation.

The flight control system can be made to be more, or less, responsive to control inputs from the front seat stick. Wilson introduced a "sloppy stick" to the front cockpit. I pulled aft on the stick until it was almost full aft, without receiving much elevator movement. At that point, pressure on the full aft stick gave additional elevator control. The same stick movement in the roll axis gave the same unresponsive control reaction.

Wilson then demonstrated the range of wing camber that can be simulated in the NF-16D by changing the flap settings. He instructed me to lower the landing gear and the flaps and establish an 11-deg. angle of attack. The speed at 10,000 ft. was 155 kt. By varying the flap settings in increments to the up position, the speed gradually increased until it reached 235 kt. at the same angle of attack and with the flaps fully retracted.

I then flew the NF-16D to a touch-and-go landing back on Runway 23 at the military field. Wilson introduced programmed turbulence during the initial approach, and I found the HUD symbology moving all over as I tried to hold a steady glidepath until he removed the input. I found that the use of the center stick during the approach and landing to be closer to my own experience and with less than 10 hr. in F-16s. However, during the 1.3 hr. VISTA flight I was able to transition back and forth between the traditional F-16 side stick and the center stick without difficulty. The only impression was that the side stick was slightly more sensitive to inputs, because of its limited movement.

The VISTA should be an excellent training aid for pilots evaluating different aerodynamic modes. It provides realistic flight visual cues and real pilot stress levels. Airborne flight accelerations also are experienced in the aircraft.

As good as simulators have become in recent years, there are still valid reasons for wanting to perform the variable stability mission in an aircraft. This is even more true when using the VISTA as a research tool to develop flight control systems.

The VISTA is equipped with an airborne test instrumentation system (ATIS) to provide on-board data recording and realtime telemetry capability. The ability to load flight control software into the NF16D for real-time evaluation would be a definite asset in the development of any new aircraft.

MATV F-16 displays high alpha benefits

David M. North/Edwards AFB, Calif.
May 2, 1994

General Electric's mature thrust-vectoring system, tested on an F110 engine, is nearly ready for production and does not require structural modifications to the airframe.

Flight testing in the Lockheed F-16 equipped with multi-axis thrust vectoring (MATV) has demonstrated that controlled high-angle-of-attack flight can be advantageous in certain tactical regimes.

Demonstrating the potential of thrust vectoring to improve tactical effectiveness in a production aircraft and engine was a primary goal of the MATV program. The project was started in 1991 by a teaming arrangement of the Air Force, General Dynamics, which has since sold the F-16 Div. to Lockheed, and General Electric. The value of thrust vectoring for takeoff and landing performance was not part of the program, because of the limited time and funding for the MATV program.

The recently completed flight test program not only proved the effectiveness of maneuvering in the post-stall regime, but also demonstrated the durability of the aircraft and General Electric F110 engine under such conditions.

"The high-angle-of-attack regime is naturally not a benefit in all tactical situations," Maj. Michael A. Gerzanics, MATV project pilot from the U.S. Air Force's 416th Test Sqdn., here, said. "But if I were in close combat with a strong opponent, I would rather have the MATV capability than not have it." Following a flight from here with Gerzanics in the F-16 MATV, I readily agree with his assessment.

During the preflight on the modified Lockheed NF-16D aircraft, Gerzanics was able to point out some unique features of the MATV aircraft. Crucial to the aircraft's effectiveness is General Electric's axisymmetric vectoring exhaust nozzle (AVEN), fitted to a baseline F110-GE-100 engine. The General Electric approach to thrust vectoring does not require structural modifications to the airframe.

Engine hydraulic oil powers three actuators, located 120 deg. apart, that in turn drive a vectoring ring. The ring is able to translate fore and aft and rotate in two degrees of freedom at the same time to provide a circular envelope of nozzle vector angles, as well as independent control of the divergent section exit area. Although the AVEN is capable of 60 deg./sec. slew rates, the flight control software limits the maximum rate to 45 deg./sec.

Also distinctive to the MATV F-16 is the spin chute, mounted above the aircraft's exhaust nozzle. The chute had to be mounted higher than normal to keep its canister out of the exhaust when the nozzle was vectored up. Ballast was installed in the nose of the aircraft to keep the center of gravity forward of 38% to preclude a deep stall if the aircraft reverted to normal control laws. The MATV F-16 had a gross weight of 28,521 lb. at the ramp.

I was to take the front seat, while Gerzanics was to fly from the back. The pilot-in-command controls, normally in the front seat of an F-16, are in the back seat of the aircraft. This includes the controls for the digital flight control system, emergency power unit, fuel system, engine mode and the environmental system. The throttle in the rear cockpit is used to start and shut down the engine, while afterburner can be selected from both cockpits.

Following completion of the prestart checklist, Gerzanics started the F110 engine and then performed additional checklist items involved in testing the flight control system and MATV controls. I then taxied to Runway 22 at Edwards and performed the takeoff.

The first demonstration was to observe the standard flight control laws of the F-16 without MATV. This was especially important as a benchmark, because my only other time in an F-16 was on a training mission with the Norwegian air force. High angle of attack performance had not been part of that flight some three years ago.

I leveled the F-16 at 28,000 ft. and slowly reached the 25 deg. angle of attack limit for the F-16 at 127 kt. With full aft stick, there was some roll control available, but rudder input had basically no effect. Pitch control was available to push the nose over and recover to level flight. The F-16 is limited to a 25-deg. angle of attack because of the loss of directional stability when the rudder loses effectiveness at near 35 deg. alpha.

The controls to engage thrust vectoring are wired in series to both cockpits. This required that Gerzanics and I both activate the switch to take the flight control system out of the "Kill" mode and into the MATV mode. The first step in the sequence to full MATV is "Standby." In this mode, the nozzle's vectoring actuators are powered, but are electrically held in the neutral position, with control laws unchanged from the baseline F-16. The next mode, "Active-limiter on" allows limited nose down pitch nozzle to aid in flight above the normal 25 deg. angle of attack limit. The mode enabling MATV flight is "Active-limiter off," which actively integrates the pitch and yaw nozzle control power to achieve controllable post-stall flight.

To add to the safety features incorporated into the F-16 MATV, the AVEN incorporates a fail-safe mode which reverts the system to a baseline F110 nozzle. The fail-safe or "Kill" mode can be commanded through a direct electrical link by the pilot or automatically by the vector electronic control. A valve in the nozzle control system hydraulically removes power from the vectoring actuators, and each actuator locks at midstroke. Electrical commands for nozzle vectoring are zeroed and disconnected from the control valve.

With active MATV, the next maneuver was an angle of attack sweep. At 28,000 ft. and a starting speed of 230 kt., I raised the nose to 30 deg. above the horizon and let the airspeed bleed off. At 110 kt. and an angle of attack of 30 deg., the afterburner was selected. The maximum amount of vectoring for the nozzle is 15 deg. in military power and 17 deg. with the afterburner selected.

However, the actual limits are a function of thrust, hardware kinematics, flow separation and aircraft loads, and are often only a percentage of the 15 deg. and 17 deg. limits. As an example, zero vectoring is allowed by the control laws at 8g, but then vectoring gradually increases to the maximum limits below 5g.

At 38 deg. of angle of attack (or alpha) there was a slight wing rock as I attempted to keep the aircraft wings level. A slight right stick input and then left rudder was required near 50 deg. alpha. No priority is given in the flight control system to either pitch or yaw. The angle of attack increased briefly to 80 deg. alpha and then settled at 75 deg. alpha. The rate of descent at this time was near 9,000 ft./min. At Gerzanics' request, I applied full left rudder, and the aircraft rolled to the left until we headed straight down and then recovered.

A similar entry was used for the next maneuver with afterburner engaged. The nose was held at a 20 deg. up attitude and 60 deg. banked turns to both sides were made at 35 deg. alpha with more than adequate roll control authority. With wings level and at close to 70 deg. alpha, left rudder was used to make a 360 deg. turn with a 35 deg./sec. yaw rate. The tracking was smooth, and there was no pitch oscillation. Stopping the nose of the aircraft after completing the 360 deg. turn was precise. Unlike most single-engine jet aircraft that require little use of rudder, I found myself depending on rudder control for flight above 25 deg. angle of attack.

Up to 25 deg. alpha, roll control laws are the same as a standard F-16. At 35 deg. alpha, roll control turns into pure stability axis roll. At 50 deg. alpha, the roll axis control is blended out so that at 60 deg. alpha there is no ability to roll the wings with lateral stick movements.

In the yaw axis, from 15 deg. to 50 deg. alpha, the rudder pedals command sideslip control and produce yaw rate with sideslip and sideslip rate feedback. At 50 deg. alpha it blends to a point so at 60 deg. alpha rudder pedals command pure body axis yaw rate with no sideslip feedback.

MATV test pilots have extended mission times during the program by using air-to-air refueling. "With four to five tanker offloads per flight, we were able to accomplish the equivalent of seven to eight unrefueled flights," Gerzanics said.

A Kansas Air National Guard Boeing KC-135 had been arranged to fuel the MATV and an F-16 that was flying chase. Gerzanics allowed me to do the initial inflight refueling, as well as another one later. As a former Navy light-attack pilot, I have done numerous refuelings using the probe and drogue system. These were my first U.S. Air Force refuelings using the boom system. After overcoming a tendency to use too much stick input for small altitude corrections, I settled into flying formation on the underside of the tanker's fuselage. The procedure was relatively easy with the boom operator doing the work.

Our next maneuver was the Cobra. Somewhat unlike the Cobra maneuver I flew with Viktor Pougachev in the Su-27 in 1990, the Cobra in the MATV appeared to be a much more controlled maneuver with predictable results. The normal pitch command rate allowed by the flight control in the MATV aircraft is 30 deg./sec. By depressing a paddle switch on the F-16's side-stick controller, the rate can be increased to nearly 50 deg./sec. At an altitude of 21,000 ft., afterburner was selected at a speed of 220 kt. As the speed hit 250 kt., I pulled full aft on the side-stick controller. The pitch rate was 42 deg./sec., and a pitch attitude of 110 deg. was reached with an indicated speed of zero. The angle of attack reading was close to 90 deg. I then nosed the F-16 over with more than adequate pitch control to level flight and an initial speed of 120 kt. A total of 400 ft. of altitude had been gained in the maneuver.

A J-turn maneuver was next. Gerzanics performed the first one. Although not speed dependent, I entered the maneuver at 230 kt. and selected afterburner. Once the aircraft was at a 90-deg. climb, left rudder input was introduced to give a 20 deg./sec. yaw rate at 150 kt. I had to unload the controls when heading straight down before pulling and going in the opposite direction. The 180 deg. heading change required 20 sec. with the pull to the horizon initiated at 120 kt. and the nose was stabilized on the horizon at a 70 deg. angle of attack.

The next maneuver briefed by Gerzanics for me to perform was a modified hammerhead. The entry was at 260 kt., and at 23,000 ft. I selected afterburner and started a 3g loop, although Gerzanics had me release some of the back pressure on the stick to maintain a constant pitch rate when passing through the vertical. With the F-16 inverted at 27,300 ft. and at 60 deg. alpha and close to zero airspeed, I pulled full aft on the side-stick controller. The velocity vector was now down, and the F-16 rapidly rotated around the aircraft's center of gravity and ended up upright and in a 10-deg. nose up attitude. Gerzanics said we had experienced a transit 138-deg. alpha at the top of the loop.

During these maneuvers, General Electric F110 showed no signs of a compressor stall or any other problem. Gerzanics had high praise for the F110 engine and said that throughout test flying at higher altitudes, high-angles-of-attack and at yaw rates reaching 50-deg./sec., there had been no engine problems. This was true when transiting through military and afterburner power at the extreme corners of the flight envelope, he said.

Several maneuvers followed in tracking the chase F-16. One maneuver involved following behind the other aircraft and then letting him become 50 deg. off the nose of the MATV in a turn. The nose of the MATV was pulled to track the other aircraft in angles of attack that exceeded its normal 25-deg. limit. During the flight, I referred only to the aircraft's head-up display. The normal altitude and airspeed were displayed, as well as angle of attack and side slip and yaw rate.

The next maneuver was termed the "split S to a helicopter turn." The chase F-16 was approaching us some 4,000 ft. below and traveling in the opposite direction. Once the aircraft was directly below, I rolled inverted with afterburner selected and did a split S. By this time the chase F-16 had executed a defensive hard nose down turn to the left. It required a pull of close to 5g near the bottom of the split S to gain the same altitude as the chase F-16. With the nose up and a stable angle of attack of 75 deg., I was able to use rudder to track the other F-16. At first, I overcompensated and had the nose too far ahead to effectively hit him with gunfire. After steadying the yaw rate, I was able to track the chase F-16 more smoothly.

One more angle of attack sweep was performed prior to landing at Edwards. This one was entered in the same method as the first one at the beginning of the flight. I was able to establish a steady 85 deg. alpha with full control prior to recovering. I then landed the MATV F-16 on Runway 22 at Edwards. This was the 78th flight in the MATV program. Total flight time was 1.7 hr.

The MATV system brings almost unlimited high angle of attack performance to the Lockheed F-16, one of the most aerodynamically restricted combat aircraft in the inventory with its 25-deg. alpha limitation. The MATV system lifts the restriction with many benefits and some cost and weight penalties.

Lockheed was responsible for system integration of the AVEN and engine into the airframe of the MATV aircraft.

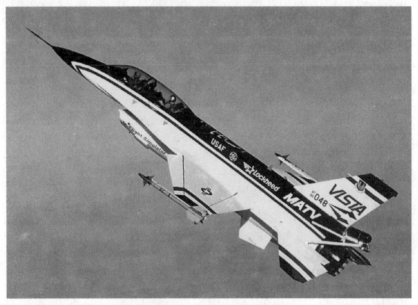

MATV flight test program concluded after 95 flights and a total of 135.7 flight hours.

After flying the MATV, I fully appreciated Gerzanics' comment that he would rather have thrust vectoring in combat than not. In a situation where pointing the nose first means winning the engagement, the MATV system would give you an edge. The combination of thrust vectoring and an off-axis, helmet-mounted display with advanced missiles would be a definite combat edge, Gerzanics said.

More importantly, in normal peacetime flying the MATV system gives a pilot an F-16 that is almost impossible to depart from controlled flight. This safety capability might have saved some aircraft and pilots' lives had the system been installed. There are no plans for installing an axisymmetric thrust vectoring system in any operational USAF aircraft.

The flight test program only investigated the benefits of MATV at altitudes of 20,000-33,000 ft. Thrust vectoring would certainly improve the F-16's takeoff performance and allow the aircraft to carry an expanded range of weapons, MATV program officials said. This potential benefit, plus the airborne capabilities of the F-16 MATV, prompted the Israeli air force to initially look at thrust vectoring. An Israeli air force pilot flew the F-16 MATV for 5 hr. earlier this year at Edwards.

Another benefit derived in part from the MATV program and other high-angle-of-attack projects is that thrust vectoring has reached maturity with the General Electric system near production capability. The fact that the system does not require structural modifications is an added benefit. As with most combat pilots, Gerzanics said higher thrust engines with MATV systems would improve the high angle of attack performance of the F-16.

One penalty imposed by the MATV system is the weight of the modified nozzle. At present the AVEN weighs about 400 lb., but GE officials predict that could be lowered to near 250 lb. The 400-lb. drag chute system would be eliminated in operational aircraft. Much of the nose ballast also could be eliminated, once testing determined the best center of gravity for an operational MATV aircraft.

"The MATV program proved that effective thrust vectoring was not engine limited, nor was it technology or aircraft limited. The final limit was of money," Gerzanics said. Gerzanics said lessons learned from the F-16 MATV program include:

- Flight research programs with goals of understanding generic high-angle-of-attack characteristics should be pursued in parallel with efforts aimed at tactical effectiveness. The MATV program emphasized potential tactical utility to explore the concept's marketability.
- Pilot selectable gain sets or control low structures should be accorded meaningful discussion during the requirements phase of flight test programs, which have some uncertainties in their model data.
- Ensure as much as possible that system ground test implementations use actual interfaces.
- Accurate air data sensing or computation is essential until control law strategies are devised which do not need to depend on air data.

Axisymmetric vectoring exhaust nozzle attached to F110-GE-100 engine provide thrust vectoring.

Nozzle durability of the F110 engine was good during flight test. Only a few flaps (lighter colored slats) and seals had to be replaced due to cracking or hot streaks.

USAF to press thrust-vectoring tests on limited budget

Bruce A. Smith/Edwards AFB, Calif.

The Air Force plans to permanently install a thrust-vectoring nozzle on the F-16 Variable-stability Inflight Simulator Test Aircraft (Vista) as part of an effort to recover from budget cuts that derailed the service's previous plans for vectored thrust research.

The Air Force had established a strategy under which it would spend $25-30 million between 1993 and 1997 for three flight test programs. Participating contractors would have invested a significant amount of their own funds in the planned research efforts.

The broad-based Air Force plan called for taking data from the F-16 Multi-Axis Thrust Vectoring (MATV) program on the Vista aircraft and combining it with results from the planned Advanced Control Technology for Integrated Vehicles (Active) project (*AW&ST* July 5, 1993, p. 30).

The experience from those programs was to be used by an all-envelope flight test aircraft later in the decade. But that program—called Propulsion, Aerodynamics, Controls Integration Research and Development (PACIR)—was killed by Congressional budget cuts for this fiscal year.

The aircraft, probably an F-15 or an F-16, would have been capable of conducting a full spectrum of thrust vector research activities, ranging from air-to-air combat and air-to-ground delivery to cruise performance and takeoffs and landings.

In addition, the Air Force's share of the NASA/Air Force/McDonnell Douglas/Pratt & Whitney Active program funding for this year was cut in Congress.

Les Small, MATV program manager at the Air Force's Wright Laboratory, said the service currently is selecting key elements of technology that are considered most relevant to future aircraft, including technology programs such as Joint Advanced Strike Technology (JAST). Those high-priority research efforts will be flown on available testbed aircraft.

Permanent installation of a thrust vectoring nozzle on the Vista aircraft is part of the plan. The modification of the aircraft—which would include an all-envelope, thrust vectoring capability—could occur in 1997, assuming the funding holds. The Vista aircraft has been returned to its earlier configuration following completion of MATV flight testing.

Another project planned to begin next year is work toward flight of the Active F-15 in a tailless, or reduced tail, configuration. The thrust-vectoring nozzles would be used to demonstrate if the aircraft retained maneuverability across the flight envelope despite smaller tail surfaces.

The Air Force had not planned reduced tail surface research on the aircraft prior to budget cuts, but probably would have during the PACIR program.

The first step in the service's initial strategy, the MATV program, has been completed successfully. The project demonstrated the value of a production representative nozzle on an operational aircraft for close-in combat maneuvering at high-angle-of-attack and low-to-moderate speeds.

MATV was unusual for the Air Force because of its rapid pace and the fact that most of the development work on the nozzle system had already been completed. In addition, most of the investment in the program was through contractor funding.

General Electric and General Dynamics—prior to Lockheed's acquisition of the F-16 division—agreed in 1991 to flight test the Axisymmetric Vectoring Exhaust Nozzle (AVEN). The Israeli air force—considered a potential customer for the system—agreed to provide an F-16 for flight tests in Israel.

General Electric obtained the necessary export licensing for the project, but the license had a provision stating there was no guarantee the technology would be made available to Israel in production form following flight test. Ultimately, the Israelis pulled out of the project, citing budgetary reasons, and the U.S. Air Force stepped in early last year to provide flight testing at Edwards.

The Air Force has spent about $6 million on MATV, while General Electric's investment on AVEN is about $20 million, according to company officials.

The Vista aircraft was available because it was in flyable storage. MATV flight testing began in July and ended in mid-March, with 95 flights and a total of 135.7 flight test hours. The only significant downtime in the flight test program was 2.5 weeks for update of flight control software and 2.5 months for a hydraulic problem unique to the Vista aircraft, but not related to the thrust-vectoring nozzle.

The Vista F-16 was available for only a limited time for MATV, but that restriction served as a motivating factor for participants to work together and complete test objectives within the allotted time.

Kenneth L. Bonnema, chief project engineer at Wright Laboratory, said the Air Force added the MATV contract on to the agreement the service already had with Lockheed for Vista modification and flight test. That significantly reduced the time required to initiate a new contract.

Various review boards for flight test of modified aircraft were either combined or introduced to the MATV program at an early stage in an effort to familiarize them with the program and its objectives. The Air Force Flight Test Center saved program money by authorizing use of a contractor-run control room at Edwards instead of the primary facility there.

Project officials maintained two specific primary flight test objectives throughout the program: clearing the flight envelope and conducting operational utility evaluations. Inflight refueling was conducted on nearly 90% of the MATV missions, resulting in sortie times that in some cases were considerably more than 2 hr.

If test data did not match predicted values, flight testing was not necessarily halted in an attempt to resolve the differences. Decisions were made on whether to continue the flight based on whether the aircraft was controllable and handling qualities were considered acceptable.

MATV program officials maintained a communications link with the two other thrust vectoring, high-angle-of-attack programs at the base: the X-31 Enhanced Fighter Maneuverability Demonstrator and F/A-18 High Angle of Attack Research Vehicle (HARV) (*AW&ST* Apr. 18, p. 46).

Two NASA-Dryden Flight Test Research Center pilots flew the F-16 MATV. The MATV aircraft's capability to revert to standard F-16 control laws when thrust vectoring was disengaged aided the program during safety reviews. David S. Kidman, an MATV stability and control engineer at the Air Force Flight Test Center, said inflight pilot-selectable gains were considered a key reason the program was able to achieve the flight control qualities attained in a limited time.

Additional selectable gains were added when Lockheed updated the flight control computer software about two-thirds of the way through envelope expansion. The software update, which Lockheed completed within several weeks, was made due to lateral directional handling problems as the aircraft went through the 35–50 deg. angle of attack region (*AW&ST* Oct. 25, 1993, p. 27). The update enabled the military utility evaluation to be conducted without restrictions on pilot control inputs above 20,000 ft. agl.

The tactical evaluation phase of the program began Nov. 10 and was conducted over seven flying days. During that period, 17 missions were flown for a total of 21.7 hr. and the completion of about 175 engagements with two operational pilots from the 422nd Test and Evaluation Sqdn. at Nellis AFB, Nev.

The Active F-15, the former STOL/Maneuver Technology Demonstrator (S/MTD), was delivered to Dryden Flight Research Center last year after being in storage for nearly two years at McDonnell Douglas facilities in St. Louis (*AW&ST* Jan. 25, 1993, p. 67). Dryden is the responsible flight test organization for Active.

The highly modified F-15E, which was used to validate two-dimensional thrust vectoring/reversing nozzle technologies in tests that ended in 1991, will be equipped with Pratt & Whitney's pitch/yaw balanced beam nozzles. First flight with vectoring nozzles is scheduled for 1995.

The project initially will evaluate the operability and performance of the thrust vectoring nozzles, and then begin study of the use of thrust vectoring to reduce trim drag at cruise conditions, including supersonic cruise.

Ronald J. Ray, Dryden chief engineer for Active, said initial simulation of tailless configurations could begin in 1995. An initial step likely would be a reduction in the current tail's size. The X-31 program also has been looking into flight with tailless configurations (*AW&ST* Mar. 28, p. 25). Ray said NASA is committed to funding and supporting the aircraft for at least the next two to three years.

SR-71 impressive
in high-speed regime

Robert R. Ropelewski/Beale AFB, Calif.
May 18, 1981

Continued improvements to aircraft systems as well as reconnaissance and electronic warfare subsystems in the U.S. Air Force/Lockheed SR-71 are keeping that aircraft a viable and survivable strategic reconnaissance platform after more than 16 years in service.

The SR-71 remains unmatched in its sustained speed and altitude performance despite its aging subsystems and the heavy workload imposed on the crews who operate the Mach 3-plus aircraft. Efforts are underway to update the SR-71's flight controls and displays with newer technology that will improve reliability and reduce crew workload.

The impressive performance and demanding workload of the SR-71 were experienced first-hand by this *Aviation Week & Space Technology* pilot in a recent flight in the aircraft with the 1st Strategic Reconnaissance Sqdn. of the 9th Strategic Reconnaissance Wing here at Beale AFB. The flight was preceded by two days of briefings and preparation, including a full day in both the front and rear cockpits of an SR-71 simulator.

It was barely enough preparation. The unique high speed, high altitude, high temperature environment in which the SR-71 operates makes for preflight, in-flight and post-flight procedures that are more lengthy and complex than any other aircraft in the Free World inventory. Special life-sustaining support systems are essential before, during and after every flight, and normal and emergency procedure checklists used by SR-71 pilots and reconnaissance systems operators (RSO) are thicker than any aircraft checklist I have seen in the past.

Despite this, the SR-71 makes flight at Mach 3.0 and 80,000 ft. seem easy. It was not until 1976, in fact—more than 10 years after the SR-71 entered service—that 9th Strategic Reconnaissance Wing crews from Beale set world absolute and world class speed and altitude records of 2,189 mph. and 86,000 ft. in horizontal flight (AW&ST Aug. 2, 1976, p. 27). These records still stand.

Crew preparations for the *Aviation Week* demonstration flight began more than three hours before the flight with a high-protein breakfast of steak and eggs in the Physiological Support Div. dining room, which is maintained specifically to meet the special requirements of the SR-71 and U-2 missions.

Mission simulation

The day before the flight, typical missions were flown from both the front and rear cockpits in the SR-71 simulator, under the supervision of Maj. B.C. Thomas and Maj. William Keller—instructor pilot and instructor RSO, respectively, in the simulator and the SR-71.

There are no flight controls per se in the rear cockpit of the standard SR-71A, although the RSO can control the horizontal flight path of the aircraft through the astro-inertial navigation system.

Neither the layout nor instrumentation in either cockpit is particularly exotic. A possible exception is the navigation system controls in the rear and the instrumentation in the front cockpit for monitoring and controlling the aircraft's center of gravity and the engine inlet and exhaust systems that provide much of the aircraft's cruise thrust.

B.C. Thomas, who had provided front cockpit simulator instruction, was the pilot for the demonstration flight, while I occupied the RSO position in the rear cockpit.

About 1.5 hr. before takeoff, we went through a brief physical examination, including temperature and blood pressure checks (standard procedure before each SR-71 mission), then began suiting up in full-pressure suits.

Two men are needed to assist the crewmembers in donning this suit, and this service was provided by personnel from the 9th SRW's

Special life-sustaining support systems are essential before, during and after every SR-71 flight. Normal and emergency-procedure checklists used by the pilots and reconnaissance systems operators are thicker than most aircraft checklists.

Physiological Support Div. The process took about 30 min., including a final check of the pressurization, breathing, cooling and face-plate heating functions.

We were then transported by van to the aircraft, with hoses available in the van for cooling the inside of the suit. For the transition from the van to the aircraft, portable cooling units were carried.

Each SR-71 is kept in its own individual shelter at Beale, and all preflight checks and engine startup are accomplished in the shelter. The shelter floor beneath the aircraft was covered with fuel, and it flowed at some points to other areas of the hangar. The SR-71 has six fuel tanks in its fuselage and wings, but there are no internal fuel bladders as such and the skin of the aircraft serves as the outer wall of the tanks.

Because of airframe expansion and contraction associated with heating of the SR-71's skin at high Mach flight and subsequent ground cooling, it is impossible to keep the fuel tanks sealed (AW&ST June 16, 1980, p. 166). As a result, the tanks leak profusely on the ground, leaving a large pool of fuel beneath and around the aircraft. "It's enough to give a safety officer a heart attack the first time he sees this," Thomas said as we walked around the aircraft.

The SR-71 is equipped with an elaborate three-axis, eight-channel stability augmentation system that automatically compensates for many of the aircraft's natural instabilities.

Lockheed SR-71 is powered by two Pratt & Whitney J58 turbo ramjet engines and is capable of speeds in excess of Mach 3.0.

Because of the fire risks with standard fuels under these conditions, JP-7 fuel with a high flash point is used in the SR-71. There is no evaporation with JP-7, and a lighted match can be dropped in it without igniting it.

Strapping into the cockpit of the SR-71 with a full pressure suit is a cumbersome process, and Physiological Support Div. personnel rou-

Majority of SR-71s were flown by a pilot and a reconnaissance systems officer; however, the SR-71B had a second cockpit installed behind the original cockpit.

tinely accompany crewmembers to assist in this. Once in, we sealed our helmets and began breathing 100% oxygen.

Standard practice in the SR-71 is to breathe pure oxygen for at least 30 min. prior to takeoff to eliminate nitrogen in the body, thus reducing the possibility of decompression sickness at altitude. SR-71 crews normally try to be in the aircraft 50-55 min. before their scheduled takeoff time, with the engines started about 40 min. prior to takeoff.

Cockpit entered

Our installation in the cockpit was one of the final steps in a process that had begun hours before for the various support personnel associated with the aircraft. For a typical operational mission, the process begins at least a day ahead. An 18-24-hr. lead time is normally needed to identify, prepare and install the reconnaissance sensors that will be used on a particular mission.

In addition, a mission tape is cut to be installed the day before the mission in the aircraft's astro-inertial navigation system. The tape provides navigation commands to the SR-71's autopilot during the flight and automatically starts and stops sensors and their recorders when the aircraft approaches and passes the ground positions designated for reconnaissance.

With sensors installed and the navigation system programmed, mission payload specialists then begin a checkout of the system about 2.5 hr. before flight. Shortly after this, Physiological Support Div. personnel begin their own checkout of environmental control and life support systems.

Lockheed SR-71 lands at Dulles International Airport. First production SR-71A flew in December 1964.

Because of the environmental extremes in which the SR-71 operates, a very thick, specially developed oil is used in the aircraft's Pratt & Whitney J58 engines. When temperatures drop below 30C (86F), the oil is almost a solid and must be preheated to 30C before the engines can be started. Normally, this takes about 1 hr. for each 10 deg. the oil must be heated. Special ground carts are required for this task, and these were operating as we climbed into the aircraft because the temperature was about 15C (60F).

Prestart checks in both cockpits took 15 min. Because of the high flash point of JP-7 fuel, normal igniters are incapable of igniting it. Instead, a small amount of triethyl borane (TEB) is injected into the engine combustion chambers using an air turbine starter plugged into the bottom of each engine nacelle once engine rpm has built up. The triethyl borane ignites on contact with the JP-7 and causes the fuel to ignite as well. The TEB also is used for each afterburner ignition and for engine air starts (AW&ST May 10, 1976, p. 93).

Once the first engine was started, it was necessary to wait 2 min. to insure that the engine's hydraulics were operating properly before any hydraulically actuated systems or flight controls could be operated. Controls were then checked and the second engine was started. Additional checklist items took another 15 min. before we were ready to taxi.

Brake malfunction

As the aircraft rolled out, Thomas applied the brakes only to find that the pedals went to the floor with no brake response. A quick recycling of the hydraulic system selector switch in the front cockpit brought pressure, however, and we continued taxiing toward the runway. Our flight plan called for an unrefueled mission that took us from central California over Nevada and Idaho, then back across Idaho and over Oregon before returning to Beale.

Although it was partly cloudy as we taxied to the runway, I selected the astronavigation function on the astro-inertial navigation system in the rear cockpit. Within a few seconds, a white star illuminated on the mode selector button, indicating that the system had located and was tracking at least three stars in its preprogrammed catalog of 52 stars.

A chronometer in the navigation system is programmed with the day of the year and the time of the day, accurate to 5 millisec., and thus knows which stars to look for from any location and at any given time. A star tracker mounted on gimbals takes periodic sightings that are then fed into a digital computer to correct the aircraft's position as determined by the system's inertial reference unit.

The navigation system, mounted in a space in the upper fuselage just behind the RSO's cockpit, thus provides the reconnaissance aircraft with precise navigation without relying on any external radio emission. Had clouds prevented the star tracker from getting an accurate fix while we were still on the ground, the inertial system would have continued to provide navigation guidance until we were above the clouds and star tracking was possible.

Remaining items on the takeoff checklist were completed in the run-up area beside the runway. Thomas held the brakes and ran the throttles up to military power (maximum power without afterburner) on one engine at a time for an engine trim check. Unlike most other turbine engines, the Pratt & Whitney J58 can be trimmed from the cockpit of the aircraft—either automatically or manually.

With the trim switches in the automatic position during the run-up, Thomas monitored the exhaust gas temperature/rpm relationship on each engine to ensure that it conformed with the figures on his checklist. EGT was monitored throughout the flight to make sure there was no significant variation between the two engines.

Once this and the remaining checklist items were completed, we taxied onto the departure runway. Thomas advanced the throttles to military power but held the brakes until getting an indication that the movable inlet guide vanes on the J58s had shifted from the axial (full open) position to the cambered position (partially closed, to turn incoming airstream at the compressor face). He then released the brakes and pushed the throttles into the afterburner range. The left burner ignited about one-half sec. ahead of the right one, resulting in a snap to the right, noticeable in the cockpit. An ignition lag of up to 5 sec. is acceptable between the two afterburners.

Aircraft weight at that point was in excess of 100,000 lb., including about 50,000 lb. of fuel. With the two J58 engines producing a combined total of about 68,000 lb. thrust, the aircraft accelerated quickly. At 180 kt., Thomas raised the nose of the aircraft to a deck angle of about 15 deg., and the SR-71 lifted off at about 210 kt. after a takeoff roll that took about 20 sec. and covered approximately 4,300 ft.

Landing gear limitation

The SR-71 has a 300-kt. landing gear extension limitation, and the aircraft reaches that speed quickly after liftoff. The landing gear retraction cycle, on the other hand, seems to be rather slow, sometimes necessitating a power reduction to keep the airspeed below 300 kt. until all gear units are retracted fully.

I had experienced this in the simulator the previous day. Thomas faced the same problem on takeoff in the aircraft, and he was just about to reduce power as the airspeed nudged 300 kt., when the landing gear finally gave a full-up indication.

Although an aerial refueling was not included in our flight, we leveled off at 25,000 ft. as if we were going to refuel from a tanker. At 400 kt. equivalent airspeed (KEAS) the climb to 25,000 ft. was accomplished quickly, with the aircraft climbing at a rate of approximately 10,000 fpm. We leveled off at that altitude, holding 400 kt.

While we waited for clearance to a higher altitude, Thomas pulsed the controls with the stability augmentation system (SAS) on and off. Pitch, roll and yaw oscillations were dampened almost immediately with the SAS on, and the aircraft showed no unfavorable tendency to diverge from its normal attitude. With the SAS off, the pulse-induced oscillations continued for at least five to six cycles before slowly damping out.

Simulated failure

In the simulator, I had experienced an SAS failure while cruising on instruments at Mach 3.0 and 80,000 ft. Chasing the nose of the aircraft with the stick under these circumstances resulted in a pilot-induced oscillation that ultimately led to 90-deg. nose pitch-up and loss of control of the aircraft.

Because of the reduced pitch stability margin at high speeds, a pitch boundary indicator in the form of an index on the left side of the indicator is incorporated in the front cockpit. The same index is used on many other aircraft to present raw instrument landing system (ILS) glideslope data. The pitch boundary indicator is controlled by both angle of attack and pitch rate inputs. In the event of stability augmentation system malfunction, the pitch boundary indicator and the pitch steering bar on the flight director attitude indicator become the primary pitch references. A stickshaker and stickpusher are also associated with the pitch control system.

In the roll and yaw axes, surface limiter switches are incorporated to prevent excessive control travel. Roll limits are applied at the stick itself, while yaw controls are imposed at the rudder servos. The surface limiters are manually engaged through switches in the cockpit that are turned on at 0.5 Mach during acceleration.

Stability system

Because of the pitch and yaw instabilities, the SR-71 is equipped with an elaborate three-axis, eight-channel stability augmentation system that automatically compensates for many of the aircraft's natural instabilities.

At the same time, however, the SR-71 is limited to a conservative bank angle even with the SAS engaged. This can be increased slightly with the autopilot engaged. Until they have logged 60 hr. in the air-craft, SR-71 pilots are restricted to shallow bank angles.

After several minutes at 25,000 ft., we were cleared to continue our climb and accelerate to Mach 3.0. Thomas advanced the throttles into the afterburner range, and there was a sudden slight yaw to the right as the afterburners ignited asymmetrically again. Pulling the nose up about 10 deg., Thomas kept our speed at 0.9 Mach until we passed 30,000 ft., then adjusted the aircraft attitude for a slow accel-eration to 0.95 Mach at 33,000 ft. Our rate of climb at that point was over 6,000 fpm.

Because the highest drag, and therefore the greatest fuel consump-tion, is in the transonic regime, SR-71 crews try to get through it as quickly as possible by performing a "dipsy" maneuver. Just before reaching 33,000 ft., Thomas pushed the nose down slightly to pass through Mach 1.0 and accelerate to 450 kt. at a steady descent rate of 2,500 fpm. He then pulled the nose up to continue the climb at 450 kt. as the Mach number slowly increased. There was no sensation or un-usual aircraft reaction as it passed through Mach 1.0.

From this point on, Thomas was busy monitoring the various fuel, engine and engine inlet systems used on the aircraft to control its center of gravity, keep the engines running at maximum efficiency and increase the thrust of the propulsion system.

The fuel system had pumped fuel aft automatically during the acceleration to supersonic speed to shift the center of gravity rear-ward in order to balance the shift in aerodynamic forces. Because the aircraft cannot be flown to its maximum cruise speed without complete control of the center of gravity, two manual backup sys-tems are incorporated to accomplish this function in case the auto-matic controls do not work. Thomas said, "More often than not, we have to play with it."

The SR-71 uses mixed compression, axially symmetric inlets and a free-floating exhaust nozzle system to control airflow in and out of the engines. The functioning of these systems is such that a turbo-ramjet phenomenon occurs in which the inlets and exhaust nozzles produce most of the thrust at cruise speeds.

At Mach 1.4, bypass doors around the forward portion of the en-gine nacelles began automatically to modulate the flow of air to the engine compressors. Forward and aft bypass doors are located in front of the compressors, with the forward doors spilling inlet air overboard when necessary and the aft bypass doors venting excess inlet air around the engine and into the exhaust nozzle, where it pro-duces more thrust.

At Mach 1.7, Thomas selected the "A" position for the aft bypass doors, opening them slightly. Also at Mach 1.7, the "spikes" or cones protruding from the center of the nacelles begin retracting into the nacelles to control the shock wave that forms in front of the engine compressors when the aircraft is at supersonic speeds. The spikes move aft when the aircraft reaches its maximum cruise speed.

Shock wave

Between them, the spikes and the forward and aft bypass doors keep the shock wave trapped inside the nacelle and ahead of the compressor, allowing air to enter the compressor at subsonic speeds.

At Mach 2.0-2.1, the variable position inlet guide vanes on the engines translate from the axial to the cambered position to maintain a constant inlet pressure on the compressor face. The pilot must engage a lockout switch at that point to prevent the inlet guide vanes from returning to the axial position. Without the IGV shift, speed is limited to about Mach 2.0.

At Mach 2.6 and 450 KEAS, the autopilot, if engaged, begins to follow an airspeed bleed schedule that reduces equivalent airspeed by 1 kt. for each incremental increase of 0.01 Mach.

By Mach 2.7, the aft bypass doors in the inlets were nearly fully closed again to meet the ram air requirements of the engines at that speed.

All of these steps have to be either monitored or accomplished manually by the pilot. Given the rate at which the SR-71 accelerates, this means that the pilot is extremely busy during the acceleration to cruise speed and the climb to altitude. As we passed 70,000 ft. at Mach 2.99, the SR-71's climb rate was still above 1,000 fpm.

When the inlet and exhaust systems are working properly, they produce up to 90% of the SR-71's thrust at cruise speeds, according to Ben Rich, Lockheed-California Co.'s vice president of advanced development projects (Skunk Works). Rich was one of the designers of the SR-71 and its complex propulsion system.

High-altitude performance

"At high altitude, the engine becomes a supercharged ramjet," Rich said. "At cruise speeds, 60% of the thrust comes from the inlets, which act as superchargers in front of the engines. The engine is simply an air inducer at that point."

In the exhaust nozzle, Rich said, air comes out of the turbojets at essentially the same speed it went in. "We designed the ejectors with convergent/divergent nozzles that reaccelerate the air to as close to cruise Mach as possible. The aircraft gets 30% of its cruise thrust from the ejectors. That means the engines are producing only 10% of cruise thrust."

Because of this, a malfunction or failure in the inlet or exhaust systems has the same effect as the loss of an engine on an aircraft with wing-mounted engines. An inlet "unstart," where the shock wave becomes unstable and is expelled from an inlet, is at least as dramatic as an engine loss on any twin-engine aircraft.

Drag increase

When one of the inlets malfunctions and the thrust it produces is lost, there is an immediate sharp increase in drag that causes the aircraft to yaw briskly in the direction of the malfunctioning inlet. I had experienced an unstart in the SR-71 simulator, but there the sudden yaw was restrained to a mild "kick" by the simulator. In the aircraft, however, the reaction can be severe enough to dash the crew's helmets against the canopy with considerable force.

Although the SAS and autopilot take care of the immediate need to apply corrective control inputs, an unstart is still a two-page emergency checklist event in the pilot's handbook because of the large number of subsystems that can cause an unstart.

Recovery must be made fairly quickly to avoid having to descend from altitude and abort the mission. Provisions are included in the propulsion control system for a sympathetic unstart and restart of the other engine to restore symmetry to the aircraft's flight controls.

During our acceleration and climb, a low rumble or buzzing developed around Mach 2.2 that could be heard and felt in the airframe. Thomas said it suggested the onset of an inlet unstart, and he suggested bracing for such an event. There was, however, no indication on the propulsion system instruments of a malfunction, and the buzzing subsided after 1-2 min. Rich suggested that the buzzing was caused by the boundary layer separating and reattaching in one of the inlets, which sometimes happens between Mach 2.2 and 2.5.

Shortly after this, we leveled off just above 80,000 ft. and at a speed slightly in excess of Mach 3.0, with an airspeed of around 330 KEAS. Thomas brought the throttles back to the minimum afterburner range, where they remained for the cruise portion of our flight.

The center of gravity at this point was at 25% mean aerodynamic chord, compared with 19% at subsonic speed. The cg. shift was accomplished automatically during the climb and acceleration phase.

At our cruise altitude, the curvature of the Earth was readily apparent, and the sky above was a very dark blue. Despite the thin air, there was enough friction on the aircraft to generate a substantial amount of heat.

Surface temperatures range from 400 to 1,200F at various spots on the aircraft's exterior during prolonged flight at Mach 3.0. Around the cockpit itself, the temperature reaches about 530F. This heat could be felt through the narrow windows of the rear cockpit, even while wearing the relatively thick pressure suit gloves. Temperatures on the wing and fuselage skin that forms the outer wall of the aircraft's fuel tanks get even hotter than they do around the cockpit, and a nitrogen inerting system is used to reduce the possibility of fire in the tanks. The nitrogen is injected directly into the fuel cells from Dewars carried inside the aircraft.

While the sensation of speed is generally lacking in most high-flying jet aircraft, the cruise speed of the SR-71 is such that its rapid movement was apparent even from 80,000 ft. when judged by the broken cloud layer 50,000-60,000 ft. below us. Groundspeed at that point was above 30 mi./min. Thomas had engaged the autopilot when we began our climb out of 25,000 ft., and that, coupled with the astronavigation system, held us closely to our preplanned course.

Accuracy checks

Within its capabilities, the system checks its own accuracy and displays the deviation from course in one of the windows of the ANS control head in the rear cockpit. The deviation is presented in tenths of a mile right or left of course. Except when the aircraft was being hand flown, the course error window generally showed all zeros. In the front cockpit, course deviations are indicated by deflections of the course deviation bar on the attitude director indicator. Full deflection of the bar either right or left indicates the aircraft is 1.0 mi. off course.

"We start to get nervous when we're just a little off course," Thomas said. "The ANS generally flies the black line," he added, referring to the system's ability to navigate accurately a course drawn with a black line on a map.

Thomas and other SR-71 crewmembers said it is possible to get lost quickly at Mach 3.0, and so careful attention is paid to the aircraft's position throughout each flight.

Fuel requirements are a major factor in mission planning and are based on the assumption that the aircraft will follow the designated course. There is little margin for error. Thus, even when getting lost is not a consideration, any deviations from the planned course can add significantly to fuel consumption and jeopardize the successful completion of the mission.

Flight plan

Pilot and RSO in the SR-71 both carry a detailed flight plan for each mission that lists normal flight planning information such as en route times and estimated arrival times over en route navigation fixes as well as specific bank angles to be used and the schedule for the operation of the various reconnaissance systems on the aircraft. The RSO monitors all of these and checks them against the estimated en route and arrival times, waypoint information, groundspeed and course deviation indications provided by the astro-inertial navigation system. All times on the plan are listed in minutes and seconds.

Other factors besides course deviation can affect fuel consumption, and the crew's monitoring function is continuous. The inlet spike for each engine, for example, is controlled automatically by an air data computer according to the aircraft's speed. If the spikes are as little as one-half in. out of position for a given speed, fuel flow could increase significantly enough to necessitate aborting the mission. Spike position indicators in the front cockpit are monitored by the pilot throughout the flight, and the spikes can be positioned manually by the pilot if the automatic system malfunctions.

Likewise, center of gravity control is also critical to fuel efficiency in the SR-71. Rather than using the elevons to trim the delta-wing aircraft, fuel is pumped forward or aft to shift the center of gravity, allowing the elevons to be kept in a minimum drag position. A 1% error in center of gravity location from the optimum can result in a mission abort for low fuel, Thomas said. "It's really incumbent on the pilot to take a great deal of care in how the aircraft is flown," he said.

As we passed to the south of Boise, Idaho, the aircraft banked to the left in a programmed 180-deg. turn that took us to the east and then to the north of the city before the turn was completed. At Mach 3.0, the SR-71 does not turn very quickly, and at a bank angle of 30 deg., our 180-deg. turn described a semicircle around Boise with a diameter of about 170 mi.

The turn took long enough to allow me to sample the specially packaged foods carried by SR-71 crews on extended missions. A soft plastic bottle with water, a tube of apple sauce and a tube of pureed peaches had been stowed in the rear cockpit before the flight. A special cap with a relatively thick plastic straw comes with each tube, and the water bottle also has an integral plastic straw in the cap.

These straws are inserted through a small hole at the base of the pressure suit helmet, then pushed into the mouth. The water bottle or the tubes are then squeezed to take nourishment. The technique worked well enough, although the system does not make provisions for inserting a napkin to wipe the peach residue off the crewmember's chin and off the inside of his mask. The hole at the base of the helmet was self-sealing once the straws were removed.

Shortly after the aircraft completed the turn and was heading back toward the west, Thomas retarded the throttles to military power from the minimum afterburner position and we began a descent. En-

gine and inlet limitations at high speed and high altitude are such that
the SR-71 has a very narrow descent "throat" through which it must
be flown when leaving cruise altitude. The engines are kept at military power—maximum power without afterburners—and airspeed is
kept at 350 kt. minimum to avoid inlet disturbances.

"The throttle schedule is locked in once we start our descent,"
Thomas said. "Mach 1.3 is the first point in our descent and deceleration where we can do anything to adjust the descent profile."

Descent rate

The descent/deceleration profile began moderately with a descent
rate of about 400 fpm and a slight deceleration rate. Our speed was
still at Mach 2.5 at 70,000 ft. and Mach 2.0 at 60,000 ft. Because the
precise altitude capabilities of the SR-71 are still held secret, common
procedure for Air Force crews is to turn off the Mode C altitude reporting function of the aircraft's transponder above 60,000 ft. I had
done this on our climb to altitude, and conversely, turned it back on
again as we descended below that altitude on our return. The rate of
descent had reached 3,500 fpm by 60,000 ft. and was at 5,000 fpm
when we passed through 50,000 ft. at about Mach 1.6.

At around 31,000 ft., the aircraft's speed dropped below Mach 1.0,
and the nose pitched up slightly from a 15-deg. nosedown attitude to
about 5 deg. nose down. Our equivalent airspeed at that point was
still about 365 kt.

Throughout the descent, Thomas had been monitoring center of
gravity, inlet spike and bypass door movements, and engine behav-
ior in approximately the reverse order from our earlier climb and acceleration. Once we had slowed to subsonic speed, the workload
decreased and it was possible to fly the SR-71 more like a conventional aircraft.

With the autopilot still engaged, I selected waypoint No. 10 in the
ANS—the coordinates for Beale AFB—and pushed the "direct steer"
button on the ANS control head.

The aircraft turned from its westerly heading to a more southerly
one, heading directly toward Beale, which was still about 80 mi. away.
Had Thomas been flying manually, he could have followed the ANS-
generated flight director steering commands on his attitude director indicator to establish the proper heading for the return to Beale.

While passing through a layer of broken clouds between 15,000 and
10,000 ft. during our descent, some light turbulence was encountered
that demonstrated the structural flexibility of the SR-71.

The center of gravity of the aircraft is located in the same general
location as that of the landing gear, roughly at the midpoint of the air-

craft's 107.4-ft. overall length. The cockpit is approximately 50 ft. in front of this.

Light turbulence

In light turbulence, the effect in the cockpit was like sitting at the end of a diving board, bouncing up and down several times during each patch of turbulence. It was not particularly uncomfortable. SR-71 crews have seen significantly more pronounced oscillations in more severe turbulence, according to Thomas.

Once in the landing pattern at Beale, the SR-71 behaved like a conventional aircraft, and even showed a surprising ease of handling that belied its size.

Thomas flew a tight pattern to an initial low pass, an approach and go-around with a simulated engine failure, a touch-and-go landing and a final landing. Airspeed was kept around 250 kt. in the pattern, slowing to 180 kt. on final.

For the engine-out demonstration, Thomas brought the right engine to idle. There was little yaw associated with the loss of thrust on the right wing because the stability augmentation system automatically applied corrective rudder inputs. The left wing was also lowered about 10 deg. to minimize drag.

The aircraft was much lighter now, and less than military power was needed on the left engine to maintain speed and altitude in the pattern. For the go-around in this condition, the left afterburner was selected at about 300 ft., necessitating a steeper left bank of about 20 deg. to maintain our track down the runway. The SAS again took care of corrective rudder inputs, and Thomas kept his feet on the cockpit floor during the climbout and downwind turn.

A conventional touch-and-go and then a final landing followed. Although the long fuselage and large rudders of the SR-71 can pose some difficulties in high crosswinds, crewmembers said the aircraft is not generally a difficult one to land. Our own landings appeared to be fairly routine.

A nose-up attitude of about 8 deg. was held on the downwind leg, increasing to 9 deg. on base, 10 deg. on final and about 12 deg. for touchdown. Forward and peripheral visibility remains good throughout the approach.

The delta wing of the SR-71 generates a considerable amount of ground effect as it nears the runway, helping to soften the landing. Both our touch-and-go and fullstop landing were exceptionally smooth.

Once we were on the ground, Thomas deployed the drag chute that is towed in the upper rear fuselage between the canted rudders.

This provides an approximate 0.5g deceleration force, and brought us forward slightly in our seats. The chute must be released from the aircraft before slowing below 60 kt. in order to avoid getting it tangled in the rudders. This was done, and we slowed to taxi speed with a considerable amount of runway remaining.

Our total flight time was 1.4 hr., during which we covered about 1,800 mi. as well as four circuits of the landing pattern.

From the pilot's point of view, there is a glaring paradox in the SR-71 in terms of the late 1950s/early 1960s technology used in the cockpit to manage a system whose performance is still considered advanced in the early 1980s.

Improvement programs

Recognizing this, the Air Force has several improvement programs under way to modernize flight controls and systems in the aircraft, with the aim of improving reliability and thereby reducing cockpit workload for SR-71 crewmembers.

One of the main improvements will be a digital automatic flight and inlet control system that is being flight tested and is expected to appear on operational SR-71s in August.

The new system will integrate the functions of several separate older units, including the present older-generation, central air data computer, analog air inlet computers that control the spikes and forward bypass doors, the present autopilot, stability augmentation system and automatic pitch warning system.

Role of U-2 high-altitude surveillance aircraft to be expanded

Donald E. Fink/Beale AFB, Calif.
June 16, 1980

Unique capabilities of the USAF/Lockheed U-2 high-altitude surveillance aircraft have enabled it to continue to play a key role in the Strategic Air Command's global operations despite the fact that it was designed more than 25 years ago as an aerial reconnaissance platform with a short-term limited mission.

That role will be expanded considerably next year when Lockheed delivers the first of 25 new TR-1 tactical reconnaissance versions of the U-2 to SAC. The TR-1s will be operated primarily in Europe to provide tactical commanders there with high-altitude battlefield surveillance, but they also will be capable of performing the present U-2 strategic reconnaissance missions.

Handling qualities and performance characteristics that have given the U-2 its operational longevity were evaluated here by this *Aviation Week & Space Technology* editor during a high-altitude mission in a two-seat U-2CT trainer.

The aircraft is one of two trainers operated here by the 9th Strategic Reconnaissance Wing's 99th Strategic Reconnaissance Squadron. The 9th's other operational unit, the 1st Strategic Reconnaissance Squadron, is equipped with USAF/Lockheed SR-71s.

Mission block time for the evaluation flight was slightly over 2 hr. The flight portion lasted 1.91 hr. and followed a standard training profile that included a climb to the upper portion of the aircraft's operating envelope.

Ground track for the flight formed a closed loop of about 590 mi. extending north and east from Beale to Reno, Nev., southwest over Sacramento to Oakland and San Francisco and then northeast to an initial approach fix over Paradise, Calif., 43 mi. northwest of Beale. Following descent from the high-flight portion of the mission, the aircraft was flown back to Beale for traffic pattern work that included three touch-and-go landings and a final full-stop landing.

The U-2CT trainers are conversions of earlier U-2Cs, which are smaller than the current U-2R operational aircraft. The last operational U-2C has been retired to storage. Despite the fact that the U-2CT is smaller than the U-2R and has different performance capabilities, its basic handling characteristics are similar enough to the U-2R to make it an effective trainer for pilots transitioning into the larger aircraft.

The two-seat aircraft were added to the U-2 fleet after the training accident loss rate made it evident they were needed to teach new pilots the intricacies of landing the long wingspan aircraft on its bicycle landing gear. The trainers were assembled from parts salvaged from crashed U-2Cs. Prior to this, new U-2 pilots were given intensive ground schooling, companion flight training in USAF/Lockheed T-33s and sent solo on their first ride in the U-2.

Since the aircraft combines characteristics of a high-performance jet aircraft with those of a sailplane and a heavy general aviation aircraft with a tail wheel landing gear, pilots need to develop special skills to handle it in the traffic pattern and especially to land it.

The U-2 has a marked tendency to float along the runway—a condition aggravated by wind gusts or runway thermal conditions — and has to be stalled at a height of about 1 ft. to insure the tail wheel makes solid contact on touchdown. The two small solid rubber tail wheels, linked to the rudder pedals, provide the major directional control through the landing rollout, so the tail has to be kept firmly on the runway.

Another challenge is to keep the U-2 on the runway centerline and aligned so as not to hit obstructions along the edge of the runway. At Beale, the runway and taxiway lights are recessed, but at other operating sites, U-2 pilots continually have to watch for runway edge obstructions.

Fuel is carried in integral tanks throughout the span of the long U-2 wing, so pilots have to manage carefully fuel flow and tank balancing to avoid landings with one wing considerably heavier than the other. In its stalled landing condition, the U-2 begins to run out of

A second cockpit was added to the U-2 airframe for training missions.

aileron authority and, under certain conditions, a heavy wing cannot be lifted to the level position.

Auxiliary landing gear units called Pogos are fixed manually under the outboard portions of each wing by ground crewmembers to facilitate taxiing and to keep the wings level when the aircraft is parked. The pogos are locked into underwing sockets with pins that are left in place during training flights that involve only traffic pattern work.

On operational or high-altitude training flights, the lock pins are removed by the ground crew once the aircraft has been taxied into takeoff position on the runway. As the aircraft lifts off, the pogos fall to the runway and are retrieved by the ground crew.

Vertical endplates resembling down-turned wingtips are mounted on the end of each wing. These are constructed from a scrub resistant composite material and bear the weight of a wing when the aircraft tips onto a wingtip at the end of its landing roll.

Under rare fuel balance or wind conditions, U-2 pilots can sometimes taxi off the runway balanced on the dual-wheel main gear unit and tail wheels. But the aircraft usually tips onto one of the wingtip endplates before coming to a complete stop.

The endplates also give U-2 pilots the option of dragging a wingtip if gusty crosswind conditions prevent them from keeping the nose aligned with the runway centerline. The U-2s have a crosswind component limitation of about 20 kt.

In its trainer configuration, the U-2CT has the same 80-ft. wingspan as the early U-2C operational aircraft, an overall length of 47.5 ft. and a height at the tail of 14.7 ft. (AW&ST May 10, 1976, p. 83). Engine for the U-2CT, which has a mission gross weight of over 20,000 lb., is the Pratt & Whitney J75-P-13 turbojet rated at 17,000 lb. thrust.

A second cockpit, with a full set of flight controls and instruments, was fitted into the avionics bay aft of the main cockpit on the U-2 to achieve the U-2CT trainer configuration. The rear cockpit is elevated to give the instructor pilot good forward and side visibility during training flights.

Two of the TR-1s SAC is scheduled to receive in the 1982 buy will be converted by Lockheed into TR-1B trainers, with a second cockpit added as on the U-2CT trainer.

Instructor pilot

Instructor pilot for the evaluation flight was Capt. Marsden G. Kelly, who occupied the U-2CT's front cockpit. Kelly, who has logged more than 700 hr. in U-2s during three and one-half years with the 99th, used to fly operational missions but now functions as the squadron standardization and evaluation pilot.

This editor began preparations for the flight on the previous day with briefings in the Beale Air Force Hospital's Physiological Support Div. (PSD). Fitting of a U-2 pressure suit and helmet and finally flying a simulated highflight profile to 75,000 ft. in the support division's altitude chamber completed the day-long training.

On the following day, the flight was preceded with a mission briefing by Kelly and a mid-morning breakfast of steak and eggs, a low-residue, high protein meal fed to all crews before high flights in either the U-2 or the SR-71.

Dry environment

The U-2 pressure suit environment is very dry—most of the moisture is removed from the crew oxygen supply to eliminate freezing problems at the aircraft's opertional altitudes—and pilots tend to perspire heavily in the suits, which are not well ventilated at the lower altitudes where the manual workload of flying the U-2 can cause considerable exertion.

To help combat dehydration during the flight, Kelly recommended taking at least one plastic squeeze bottle of a fortified fruit drink. This editor also took a squeeze tube of apple sauce, which is prepared and packaged in a manner similar to foods carried into space by U.S. astronauts. Tube foods and liquids are ingested in both the U-2 and the SR-71 by pushing a plastic straw through a self-sealing port in the pressure helmet neck ring.

On operational missions, which in the U-2 can last up to 11 hr., pilots take quantities of liquid and tube food. A low-residue breakfast is important before long-duration missions, because the pressure suit can be fitted only with a urine collection system.

Following breakfast, Kelly and this editor had a preflight physical examination and reported to the pressure suit room, where PSD specialists assisted with the donning and checkout of our pressure suits.

The suit used in the U-2C is a partial pressure garment with an inner restraining liner that covers the torso and upper legs. That portion of the body is ventilated by circulating air and restrained in a sealed pressure environment if the cockpit pressurization system fails or the cockpit is decompressed explosively.

The arms and legs are protected by a secondary pressure liner and are restrained by external inflatable tubes when the suit is pressurized. The hands are protected by pressurized gloves and the feet by heavy flight boots.

Pilots flying the U-2R and the SR-71 are protected by full-body pressure suits that are more comfortable—especially the latest SR-71 suit—and provide better protection at higher altitudes.

Suit-up and pressure checks routinely take about 30 min. As soon as the pressure helmet is fitted and the visor sealed, the pilots start breathing 100% oxygen. U-2 pilots are required to breathe 100% oxygen for at least 60 min. prior to takeoff to purge a maximum of nitrogen from their blood streams as a precaution against the bends. This is especially important in the U-2, because its standard operating profile involves a rapid-rate climb to altitudes above 60,000 ft.

Once the suit-up and pressure checkout processes were completed, Kelly and this editor relaxed in padded recliner chairs for the remainder of the oxygen prebreathing period. The pressure suits are cumbersome, so the pilots are made as comfortable as possible prior to high-flight missions, which are physically taxing.

Again because of the bulky suits and the need to pre-breathe oxygen, U-2 and SR71 crews are not able to perform normal preflight checkout procedures on their aircraft. A backup crew system was established under which an alternate pilot—or two-man pilot/reconnaissance systems operator crew for the SR-71—is selected for each mission. The backup crews are briefed on each mission and prepared to take the flight if the primary crew has a physical problem.

Backup pilot for the evaluation flight was Maj. Charles D. Voxland, who completed preflight checks on the aircraft and helped Kelly and this editor get settled in our cockpits. We had been transported from PSD to the aircraft in a van fitted with recliner chairs. During the transfer from PSD to the cockpit, portable oxygen units were used to maintain the preflight breathing cycle.

When Kelly had completed the few preflight tasks remaining and was ready to start the engine, Voxland shifted to mobile control officer duty and manned a highpowered automobile that preceded us to the runway. As mobile control officer, Voxland worked in conjunction with the control tower at the end of our mission and used the car to chase the aircraft down the runway, giving us radio callouts for the final few feet of the landing descent.

Engine start procedure for the U-2's J75 engine is relatively uncomplicated and uses a mobile air compressor cart. Following a check of engine instruments, Kelly called ground control for clearance to Beale's 12,000-ft. runway, which is oriented 140/320-deg.

Fuel quantity at engine start—the J75 burns JP-TS (thermally stable) fuel—was 863 gal., sufficient for a flight of 4 hr. with reserves. The fact that the fuel gauge shows the quantity in gallons instead of pounds is an indication of how old the U-2 design is, Kelly said.

The aircraft does not have a fuel flow meter other than the numerical gallon indicator, which would have to be timed, so fuel status is plotted on a graph against mission time. Oxygen pressure and quantity, computed in liters for the U-2CT's high-pressure gaseous oxygen system, also is plotted on a time graph.

For takeoff, Kelly positioned the aircraft on Runway 14 at a taxiway intersection about 4,500 ft. from the threshold. Surface winds were light—140 deg. at 6 kt.—and the runway temperature was 65F. Visibility was more than 7 mi., with scattered clouds.

The fuel meter showed 840 gal. just prior to takeoff, which was equivalent to a fuel load of approximately 5,500 lb. Zero fuel weight of the U-2CT is about 14,000 lb., so gross takeoff weight was slightly over 19,500 lb.

A 6-deg. nosedown trim setting was selected, a precaution against over-rotation on liftoff, and Kelly advanced the throttle to maximum takeoff power on receipt of takeoff clearance from the tower. Brake release was at 12:24 p.m. PDT. The J75 engine takes about 8 sec. to spool up from ground idle, so the initial part of the takeoff roll was slow.

When the engine reached full thrust, however, it boosted the aircraft down the runway with a rapid acceleration approaching that of an aircraft carrier catapult launch.

The aircraft flew off the runway after a takeoff run of about 400 ft., and Kelly pulled the nose up smoothly and pegged the airspeed at 160 kt. Using the right wingtip as a reference, this editor estimated the climbout angle was approximately 60-deg.

Kelly agreed with the estimate, saying there is no way to determine the exact pitch angle, since the only attitude instrument in the cockpit is a small J-8 standby indicator that shows a maximum of 27-deg. "We use a basic cruise/climb profile, holding 160 kt. to about 52,000 ft., where we intercept the Mach schedule," he said. "From then on, we continue to climb at 0.72 Mach."

Oakland Air Traffic Control Center was working our flight—code named Hubba 55—and cleared us for an Aspen 3 standard departure from Beale and a "direct climb to above flight level 600." Air Force limits references to the U-2's operational ceiling to "above 60,000 ft."

Kelly said the clearance to climb directly to altitude would ease his workload considerably, since leveling off at intermediate altitudes in the U-2 is an imprecise as well as a time-consuming process. The imprecision results from not knowing the exact rate of climb and therefore not being able to judge where to begin leveling off. Step climbs also involve large throttle and pitch changes, which U-2 pilots find distracting, and the opening and closing of the engine intake bleed air vents to compensate for the power changes.

For high flights, U-2 pilots communicate with air traffic control on a special frequency and use a constant transponder code—in our case,

4460. This eliminates the need for changing frequencies and transponder settings, procedures that are complicated by the pressure suit.

Rate of climb during the initial portion of the cruise/climb profile also was not readable because the vertical velocity indicator was pegged. However, it could be computed by timing the climb to intermediate altitudes. At 12:28 p.m.—4 min. after brake release—our flight was passing through an altitude of 32,000 ft.

As Kelly turned the aircraft to a heading of 06-deg., following the Aspen 3 departure profile, he called attention to the fact that we had flown barely to the end of the Beale runway.

The aft cockpit in the U-2CT is small, but not as cramped as the cockpits in some jet fighters this editor has flown. The most distinctive feature is a heavy control column with a control yoke that appears more suited for a large transport cockpit.

The U-2's flight controls are operated by an unboosted cable and pulley system, and the large yoke is needed to handle the heavy control forces encountered during flight at lower altitudes and in the traffic pattern. Control forces are very light and the control response is smooth during high-altitude operations.

Basic layout of the cockpit is similar to that of most USAF jet fighters, with the throttle, flap and speed brake controls mounted on the left side and radio and navigation controls on the right side. Flight instruments are grouped on the left half of the instrument panel and engine instruments on the right.

The front cockpit of the U-2CT and the cockpits of all single-seat operational U-2s have a periscope mounted in the floor ahead of the seat. This optical system, which has magnification settings of 0.4 and 1.0, is used for observation, navigation and drift computations and defensive surveillance.

Controls unique to the high-altitude mission, primarily those regulating temperature and ventilation of the cockpit and pressure suit, are interspersed with standard equipment controls on both sides of the cockpit.

An electric deicing fan is mounted on the left side of the canopy, which swings open to the left for cockpit ingress and egress on the ground.

During the initial phase of the climbout, Kelly reminded this editor to detach the zero-delay parachute lanyard from the D-ring.

The ejection seat in the U-2CT has limited performance, compared to newer jet fighter seats. Its envelope does not include spin or dive attitudes below an altitude of 10,000 ft. or speeds below 200 kt.—in level flight only—down to an altitude of 500 ft. The U-2Rs and SR-71s have zero/zero ejection systems.

Rate-of-climb performance degraded only slightly as Kelly maintained the 160 kt. cruise/climb speed through the Aspen 3 departure profile and into a right turn to a heading of 73 deg. for the first leg to Reno. Six min. after takeoff, the aircraft was climbing through 46,000 ft.

The remaining 67 mi. of the leg to Reno were flown on autopilot. The pitch attitude dropped considerably even though the rate of climb exceeded 6,000 ft./min. through 53,000 ft., where the Mach schedule was intercepted. Rate-of-climb averaged almost 5,000 ft./min. through 60,000 ft.

At the Reno Vortac, Kelly initiated an autopilot "bug turn" to the right to a heading of 229-deg. for the 135-mi. leg to the Sacramento Vortac. A "bug turn" involves putting the autopilot on heading hold, selecting the desired heading and letting the autopilot take the aircraft to it. Autopilot turns in the U-2CT are made at 12-deg. bank angles, which in the case of the 156-deg. heading change at Reno gave a turn radius of about 20 mi.

"We have to hold our airspeed within narrow limits at altitude, so if we use higher bank angles—we can go up to 30 deg.—we lose a lot of altitude because of the reduced lift," Kelly said. "We can make tight turns if there is an operational need to do so, but normally our objective is to maintain as stable a platform as possible." Autopilot turns in the larger U-2R are made at 20-deg. bank angles, and the maximum bank angle for tighter turns is 45-deg.

During the Reno/Sacramento leg, Kelly switched off the autopilot and let this editor evaluate the aircraft's high-altitude handling qualities. Above 60,000 ft., the U-2CT is quite responsive in both pitch and roll. Pitch inputs had to be made with care, however, because the spread between stall speed and the onset of Mach buffet was about 8 kt.

The difference between Mach and stall buffet was assessed by pitching the nose down slightly and letting the airspeed approach the Mach 0.80 red line for high-altitude flight. Mach buffet started well before the limit and served as a positive warning.

Stall buffet was experienced by raising the nose and letting the airspeed drop slightly. Again the onset of buffet gave a clear warning that a critical airspeed was being approached, but this pilot could not detect a major difference between the two types of buffet.

One U-2 pilot said that during turns at operational altitude, the airspeed spread is so narrow one wing sometimes is in Mach buffet and the other in stall buffet.

Initial flight plan for the evaluation called for an autopilot right turn over the Sacramento Vortac to a heading of 353-deg., followed by the start of a descent to the Paradise initial approach fix. But the flight was going so smoothly that during the Reno/Sacramento leg it was decided to extend the mission to the San Francisco Bay area.

Kelly called Oakland Center and extended the clearance from Sacramento direct to Oakland Vortac, followed by a right turn over San Francisco Bay to a new heading of 13 deg. for the final leg to the Paradise initial approach fix.

The rate of climb had dropped considerably by the time our flight passed Sacramento, and additional gains in altitude became a function of fuel burn. Kelly maintained the J75 power setting at an engine pressure ratio (EPR) of about 3.1, slightly below the placard limits, which ranged from 3.23 to 3.27.

The U-2CT has an intermediate stop, called a gate, on its throttle quadrant that limits the engine to about 93-94% thrust. Maximum thrust can be elected by pushing the throttle through the gate to the 100% setting, as Kelly did for the takeoff from Beale.

During the high-altitude portion of our flight, the engine was operated at a setting of about 90%, which Kelly said was commonly used. A higher altitude could have been achieved with the fuel load our flight had remaining if the engine had been operated at the EPR limits, he said.

Visibility throughout the flight was excellent. Scattered clouds along the initial portion of the route served to accentuate ground features, rather than obscure them. Lake Tahoe on the eastern slopes of the Sierra Nevada Mountains was a particularly prominent landmark. At the peak altitude, the line-of-sight visibility was unrestricted and the curvature of the earth was evident along the horizon.

The sky appeared to consist of several layers of different shading, ranging from light blue where the scattered clouds met the horizon to deepening tones of blue that became almost black overhead.

The air was smooth, with no detectable turbulence, and this effect, combined with the reduced groundspeed reference at the high altitude, created an impression of floating rather than flying. The only evidence of other aircraft in the sky came from contrails jet aircraft were forming far below us.

Kelly said one of the beauties of high flight is that pilots never have to worry about other traffic.

"Our navigation on high flights is a combination of Tacan and visual reference," Kelly said. "We have the capability to use a sextent in the U-2CT, but we're not currently doing so."

A 400-kt. groundspeed is used to compute the cruise legs of a high flight mission. True airspeed on our flight averaged about 415 kt. Winds at the operating altitudes are almost always light and variable.

Over the Sacramento Vortac, a left turn was made to a heading of 210-deg. for the leg to Oakland. Outside air temperature was a constant −55F. While hand-flying the aircraft, this editor noted that the right wing felt heavy, a condition Kelly said resulted from unequal fuel flow from the wing tanks. The condition would be rechecked during the initial approach to Beale, and the wing tanks would be balanced with a transfer pump.

The autopilot, which was engaged for the remaining segments of the high flight, compensated for the fuel imbalance with a slight right aileron input. Kelly said the autopilot controls the aircraft more precisely than a pilot can and therefore is used heavily on missions that require a stabilized attitude and accurate ground tracking.

"Not only is it difficult to hand fly this aircraft all the time, it is very tiring and takes all of your attention," Kelly said. "This distracts the pilot from other aspects of the mission." Under certain conditions, however, the autopilot tends to be too sensitive to transient changes in the ambient conditions at high altitudes.

"In some cases, we encounter temperature spikes and the autopilot changes the pitch attitude to maintain its Mach hold," Kelly said. "This sometimes results in pitch oscillations because the autopilot is sensing too many variables too quickly."

If that condition is encountered, U-2 pilots usually disengage the autopilot, stabilize the aircraft manually and reengage the automatic flight control system.

High-altitude reconnaissance pilots have to train their eyes to develop a new sense of distance and size judgment, Kelly said. The change in perspective was evident during the flight over the San Francisco Bay area when this editor mistook the mothballed Liberty Ship fleet for a cluster of moored barges.

A fuel check midway through the mission showed that about 270 gal. had been burned in the first 1 hr. 5 min. Kelly pointed out that most of it was burned on takeoff and the initial climb to altitude. "We don't monitor fuel flow closely, but when computing emergency reserves, we figure we burn 187 gal. an hour at 10,000 ft.," he said. "Cruising at altitude, we burn about 160 gal. an hour and in the descent we don't use any fuel at all."

After crossing San Francisco Bay, Kelly turned to a heading of 13-deg. for the final segment to the Paradise initial approach fix. Oakland Center cleared our flight for a descent from above 60,000 ft., and this was initiated about midway along the 123 mi. leg to Paradise. The initial approach fix is a holding pattern at an altitude of 20,000 on the 331-deg. radial from the Beale Tacan.

Descent from high-altitude cruise was initiated by opening the bleed air vents in the engine intakes, lowering the landing gear and opening the speed brakes for drag and maintaining an engine pressure ratio above 2.08. In the lower portion of the descent profile, the EPR setting was reduced to 1.90 and airspeed was held at 170 kt. IAS.

A strict Mach/airspeed schedule has to be maintained during the upper portion of the descent to keep the aircraft in the narrow Mach/stall corridor, Kelly said. The engine pressure ratio schedule also has to be held to prevent a compressor stall in the J75 engine.

The U-2CT has a glide ratio of 20.6, and its best angle of glide speed is computed at 115 kt. for the average mission return gross weight.

As the descent continued, Kelly switched on the gust damping system, which deploys the flaps in a 5-deg. up position and both ailerons in a 9-deg. up position. This degrades the wing's lift and increases the rate of descent without increasing the airspeed.

"The main reason we do this is to reduce the loads on the tail by losing lift from the wing," Kelly said. The "gustsup" device also is used during level flight to reduce tail loads in turbulent or gusty conditions.

Kelly disengaged the autopilot below 30,000 ft. and let this editor hand-fly the aircraft to the holding fix at 20,000 ft. Control forces were quite heavy, and it was evident the large control yoke was needed for the manually operated flight control system. Once our flight was cleared for the initial approach to Beale, Kelly took command of the aircraft and flew the remaining 43 mi. on a straight-in approach to Runway 14.

During the initial approach, Kelly retracted the gust system and computed the threshold speed (T-speed) that would be used to establish our airspeeds for key positions in the landing pattern. T plus 30 kt. is used for initial, T plus 20 kt. for downwind and T plus 10 for final.

As a rule of thumb, U-2 pilots multiply the remaining fuel by 2, drop the last two figures and add the remaining number to what is called the aircraft's zero fuel speed. In the case of the U-2CT, the zero fuel speed is 80 kt.

Since we had 480 gal. of fuel remaining, our T-speed was 89 kt. and the initial, downwind and final approach speeds for the first landing were 119 kt., 109 kt. and 99 kt., respectively.

Once the T-speed had been computed, Kelly selected approach flaps—35 deg. is full flaps—and slowed the aircraft to 99 kt. to check the fuel balance in the wings. The aircraft tended to roll off on the right wing, so the fuel transfer pump was switched to the left position for a brief period.

Additional traffic

Other traffic operating around Beale during our first approach included an SR-71 in formation with its USAF/Northrop T-38 pace/chase aircraft, several KC-135s and another T-38 shooting touch-and-go landings. Beale tower cleared our flight for a straight-in approach to Runway 14 to meld our slow flying aircraft with the other traffic. The wind had shifted to a slight crosswind component, 160-deg., at a velocity of 6 kt.

The final approach path crosses varied terrain, including the extensive excavations of the Yuba Valley gold fields, and some small bodies of water. This generated varied ground thermal activity that was felt as pitch oscillations as the aircraft neared the ground.

Kelly selected full flaps on long final and made the first landing as this editor followed the process with light pressure on the flight controls. The need for the large control yoke and the long rudder pedal throws was evident during the final descent as Kelly made decisive rudder, elevator and aileron inputs to guide the aircraft down to the required 1-ft. height above the runway centerline.

Voxland, who was speeding along behind the aircraft in the chase car, began radioing aircraft attitude and descent rate information as soon as we crossed the runway threshold line.

When Voxland called out the 1-ft. height, Kelly held the aircraft off the runway until the onset of stall buffet was felt and then brought the control yoke back quickly to induce the full stall. While doing this, he was working the yoke through large aileron deflections to keep the wings level.

At touchdown, Kelly had the control yoke deflected to its full right roll position to keep the wings level. Large aileron and rudder inputs also were required to maintain a wings-level attitude and runway heading while the flaps were being retracted for takeoff.

When the throttle was advanced for takeoff, this editor counted to eight before the engine spooled up and delivered full thrust. This lag in engine response contributes to the difficulty of learning to land the U-2 because engine thrust cannot be used to stabilize the aircraft or prevent a stall if the aircraft skips on the initial touchdown or balloons on a gust or heat thermal.

Kelly retarded the throttle seconds after takeoff to keep the aircraft from climbing above the 1,100 ft. traffic pattern altitude used at Beale.

The second landing was made from an overhead break entry at pattern altitude. Full flaps had been selected on the downwind leg and a new T-speed computed to account for the reduced fuel load.

Following the takeoff, Kelly turned the aircraft over to this editor, who flew a standard left-hand pattern to final for the third landing.

Difficult handling

Cautions from other pilots that the U-2 "handles like a truck in the traffic pattern" were found by this editor to be only slightly exaggerated. However, it was relatively easy to stabilize the aircraft on final and cross the threshold near the desired T-speed. Voxland's callouts of the final descent were a valuable aid, since this editor found the combined tasks of keeping the wings level, maintaining runway heading and staying near the runway centerline took a major part of his concentration.

The aircraft ballooned slightly and floated with the left wingtip down before this editor managed to bring it to a fully stalled condition for touchdown. The left-wing-down attitude persisted through the initial rollout as this editor's right aileron inputs were not sufficient to compensate for the crosswind and slightly heavier fuel load in the left wing.

Kelly raised the flaps for the final takeoff, and the lag in engine response was even more evident when this editor advanced the throttle with no immediate effect.

Kelly took command of the aircraft shortly after the takeoff and made the final full stop landing.

The U-2CT has a stall speed about 11 kt. under the T-speed, Kelly said. But he added that during the final landing process, no one has time to monitor the airspeed indicator to verify that.

Gust component

When landing in gusting wind conditions, the T-speed used on final is adjusted by adding the gust component, Kelly said. Since gusts are not critical in other portions of the landing pattern, they are flown with the standard components added to the T-speed.

In some cases, especially for no-flap landings that require a faster speed on final, adding the gust factor can mean the final approach leg has to be flown faster than the downwind leg.

The key to all U-2 landings is to stabilize the aircraft as far out on the final approach leg as possible, Kelly said. This is especially true for no-flap landings that are made from a long, low-level power approach. A typical no-flap final approach begins about 2 mi. from the runway at an altitude of 250 ft. above the ground.

Chapter 4

Trainer

T-45 stable platform for advanced training

David M. North/Kingsville, Texas
August 30, 1983

The McDonnell Douglas T-45 promises to live up to the U.S. Navy's expectations of improved operational performance, a greater training capability and much lower costs than the aircraft it will replace as the service's advanced trainer.

The Goshawk, a naval derivative of the British Aerospace trainer, recently entered service with VT-21 here. While there are more than 12 Goshawks on the flight line, student pilot training will not start until next year.

This *Aviation Week & Space Technology* pilot recently had the opportunity to spend several days with the training squadron to evaluate the new pilot training system, and to fly the T-45 in a normal training syllabus flight. Since the initiation of the T-45TS program in the early 1980s, the Navy has insisted that the advanced pilot training program include all of these vital elements.

This approach dictated that equal emphasis be given to ground school, simulator training and the training aircraft. By using this approach from the start, the Navy has crafted a ground training syllabus that is integrated with the Hughes Training, Inc., simulators and the T-45.

McDonnell Douglas was scheduled to make 13 modifications to the T-45s to address the results of static fatigue tests. The changes were to include strengthening the wing.

The Goshawk is an important element of the training equation. The T-45 is slated to replace both the Rockwell International T-2 and the McDonnell Douglas TA-4J in the advanced pilot training role. The cost of operating the T-45 is projected to be approximately one sixth that of the TA-4J and one third that of the T-2. At the same time, the operational flexibility of the Goshawk is expected to reduce the amount of time required for advanced training by 15%.

I was to experience this operational flexibility firsthand by flying a T-45 with VT-21. It was decided that I would occupy the rear seat of the T-45 because of previous flights in the front seat of the Hawk trainer, and to observe the flight from the instructor's perspective. The first part of the flight was to be a four-aircraft air-to-air gunnery mission with a T-45 towing a banner. The aircraft are not yet equipped with guns. During the second half, I would be able to fly the T-45 to evaluate its handling characteristics.

The flight was with Marine Capt. Blaise Harding, a former McDonnell Douglas F/A-18 pilot and now a full-time instructor with VT-21. A flight briefing on the gunnery pattern was conducted by Cdr. W.G. (Montana) Dubois, the flight leader and chief flight instructor of the Strike Instructor Training School here.

Following the briefing, we manned the aircraft after a short pre-flight walkaround of the T-45. All of the maintenance, logistical support and ramp support for the T-45 is provided by McDonnell Douglas.

The gross weight of the Goshawk on the 100F day was close to its normal 12,750 lb. takeoff weight. The internal fuel load was at its maximum of 2,900 lb. The Navy has a modification program to add 35 gal. of fuel in tanks installed around the engine inlet. The additional fuel would allow the aircraft to taxi, take off and reach 250 kt., Navy officials estimate.

Once strapped into the back seat, and with the Rolls-Royce engine running, Harding was able to effectively cool down the cockpit on the hot afternoon flight. The fuel flow at idle for the Adour Mk. 871 engine was 600 lb./hr. Once in radio contact with the rest of the four-aircraft flight, Dubois said that he had a fuel pressure warning light and would have to join the flight later. It turned out to be a faulty sensor and was a rare occurrence in the T-45 fleet.

As a tribute to the aircraft's operational simplicity, Dubois was able to man another nearby T-45 and start taxiing with an 8 min. delay.

During the taxi to the active runway, I took control and worked the brakes and nose wheel steering. The B.F. Goodrich brakes were effective and did not grab at taxi speed. The T-45 has a digital nose wheel steering system with a high gain for close-in maneuvering and a low gain for other applications. I found that I did overcontrol the steering commands at first, but then the corrections became more gradual. While flying the T-45 simulator prior to the flight, I found the nose wheel steering to be very sensitive, especially on landing. The Navy is still evaluating nose wheel response to determine whether it needs to change the gains.

Harding initiated the takeoff as the No. 3 aircraft with a 7 sec. interval between Goshawks. I took control once the flaps and landing gear were raised and we were accelerating to rendezvous at 250 kt. Power response from the Adour engine was smooth and predictable, enabling me to make a cautious approach to the lead aircraft with a relatively slow closing rate.

While flying on the left wing of the lead aircraft, I was able to maintain a tight formation with minor control and power inputs. The Goshawk retains the positive flight control characteristics of the basic Hawk, with some improvements. Deployment of the side-mounted speed brakes brought a gradual decrease in speed at about 250 kt., but without a need for any trim change.

Harding was to fly the first gunnery patterns, while I would fly some of the later runs on the banner. As a former Navy A-4 attack pilot, I had not flown an air-to-air gunnery pattern since receiving my training here many years ago.

During the run into the pattern from alongside the T-45 tow aircraft and then to the perch position, I had more opportunity to look around the rear cockpit of the Goshawk. The analog cockpit instrumentation is functional, but the design dates to the A-4 era. The basic Hawk first flew in 1974, and deliveries began to the Royal Air Force in 1976. The U.S. Navy had initially opted for a multifunction display-equipped cockpit, but that was one of the first features to be dropped when the service was looking to cut costs in the early 1980s.

The Navy and McDonnell Douglas have initiated a development program for new instrumentation called Cockpit 21 that will feature two large multifunction displays. The No. 37 T-45 will be the prototype aircraft, and the Navy plans to put Cockpit 21 in the No. 73 aircraft and after. A three-year retrofit program is planned to outfit earlier aircraft with a cockpit more like the ones in operational squadrons.

Harding flew the first four gunnery patterns. This entailed arriving at the perch position some 3,000 ft. above and abeam the tow banner aircraft. The roll-in is started when the other three aircraft are in sight, and a smooth descending turn is made to track the banner.

Cockpit 21 upgrade with multifunction displays was to include the incorporation of the 1553B multiplex bus into the T-45.

After pulling away from the banner, the aircraft is flown parallel to the tow aircraft and to safely ahead of the tow T-45 before returning to the perch.

I flew the next three patterns and found the aircraft very stable. The roll rate of the T-45 was quick, and I would put it in the same category as the A-4. The pitch rate of the T-45 above 300 kt. appeared similar to that of the General Dynamics/Lockheed F-16B. During the turn in to the banner, there was no need for trim changes, and rudder was not needed to provide smooth tracking of the target. When the Navy first accepted the T-45, the service cited the longitudinal and lateral/directional stability as inadequate for its new advanced trainer. To correct what the Navy saw as deficiencies in the T-45, a central ventral fin was added, and both the vertical and horizontal stabilizers were lengthened.

The question of aircraft stability was not an issue during the flight. I found the Goshawk to be stable in all axes, whether tracking the banner or performing aerobatic maneuvers later in the flight. During the earlier simulator period, I flew two rocket attacks on a ground target and found that the T-45 tracked smoothly from roll-in at 8,000 ft. to firing the rockets at 3,000 ft. in a 45-deg. dive.

Visibility from the back seat while performing the gunnery pattern was very good. I was able to track the other three T-45s in the pattern without difficulty. The visibility over the front cockpit also was good, allowing me to see the upper left and right portions of the instrument panel. The ability to see the weapon select panel on the left side of the instrument panel could be helpful to an instructor.

Following a number of other gunnery patterns by Harding, we detached from the flight so I could fly the T-45 independently. We did not take the T-45 to a higher cruising altitude. Harding said that

Cockpit of a T-45 with analog technology.

at 33,000 ft., the fuel flow of the Rolls-Royce engine was 1,100 lb./hr. at Mach 0.74. He said that the TA-4J at Mach 0.8 at the same altitude would have a fuel burn of close to 2,800 lb./hr. There are no plans in the training squadron to install drop tanks on the Goshawk.

The engine quickly accelerated the T-45's speed from a low 200 kt. to 350 kt. for aerobatics. The Adour engine did not protest any of the rapid throttle changes made while checking acceleration or performing aerobatics. While the gross weight of the Hawk increased close to 1,000 lb. in the naval Goshawk, the thrust of the Adour engine also was increased. The original 5,400-lb. thrust F404-RR-400 engine was replaced by the F405-RR-401 with 5,845 lb. of thrust. Harding and other VT-21 squadron pilots said they have had no problems with the Adour engine during the more than 1,700 hr. accumulated in squadron service since December.

At an altitude of 14,500 ft., I retarded the throttle to idle to enter a clean stall. The deceleration rate was 1 kt./sec. and the stickshaker activated at a 21.5-deg. angle of attack at 118 kt. The T-45 entered a very benign stall at 110 kt. at an angle of attack of 25 deg. The Goshawk's left wing fell off slightly at 108 kt., but aileron control was effective throughout the stall and recovery.

With the landing gear and flaps extended, the stickshaker activated at 107 kt. and at an angle of attack of 21.5 deg. The stall came with a simultaneous slight left wing drop at 30 deg. angle of attack and 93 kt. By adding power and establishing a 23 deg. angle of attack, the altitude loss was less than 400 ft. Harding then kept the slats retracted during the next dirty stall routine, and the aircraft stalled at a 10 kt. higher speed.

Navy officials admit that the need for slats for the T-45's wing was marginal, but the addition gave the trainer predictable stall characteristics and lowered the approach speed to 125 kt. at a full fuel load.

The Goshawk is spin-resistant, but can be put into a spin with active control inputs. The Navy, however, is not planning to include spins in the advanced training syllabus for the T-45. The reason is that current Navy/Marine Corps fighter and attack aircraft are not spin prone, so the service is changing its philosophy.

I then flew the Goshawk back to the landing pattern at Kingsville. The Navy-type break was entered at 350 kt. with a left downwind. The speed brakes were extended to slow the Goshawk to the landing gear and flaps extend speed. Landings are made with the speed brake extended to keep engine power in a higher range for quicker response.

In another modification from the basic Hawk, the speed brakes were relocated to the sides of the rear fuselage. Later in development of the Goshawk, the Navy had McDonnell Douglas increase the throw of the speed brakes for greater effectiveness. The speed brakes are interconnected with the horizontal stabilizer, so there is little trim change when they are extended or retracted. Another change incorporated in the T-45 is the interconnection of aileron and rudder control below 217 kt.

I flew three approaches at 120 kt., with a carrier-type pattern at 500 ft. downwind. Although I am not accustomed to flying carrier landings from the rear seat, they were passable, but not "OK" No. 2 or No. 3 wire landings. The visibility was excellent from the rear seat during the landings.

The Goshawk was extremely stable during the approaches and landings, giving the pilot the feel of a larger and heavier aircraft. However, I found myself reducing too much power when rolling wings level on the straightaway, and then when correcting was high and fast. I am sure that with more time in the aircraft, this situation could be avoided. The T-45 is not the most speed stable aircraft I have flown.

On the two touch-and-go landings, I did not experience any noticeable engine lag when going to full power. The Navy is just now installing modified Lucas fuel controls on the Adour engines to eliminate a power lag that was experienced in some situations.

The total flight time was 1.3 hr. and the fuel remaining was 400 lb., for a total fuel burn of 2,500 lb. All of this flight was made below 20,000 ft. and included 14 gunnery patterns, aerobatics and three landings. The total blocks to blocks time was close to 1.5 hr.

From my experience flying a multitude of trainers, I would expect the Goshawk to be an excellent advanced trainer for Navy/Marine Corps pilots. It is an easy aircraft to fly, but not so easy that a student would become complacent.

It is a forgiving aircraft, so in order to get into real trouble, a pilot would have to persist at fouling up. British Aerospace stressed simplicity in designing the Hawk's systems, and that philosophy has been carried over in the Goshawk. The good safety record of the Hawk and the proven reliability of the Adour engine also should enable the Navy to have a low T-45 accident rate.

Following the flight, Lt. Tim Smith, a VT-21 landing signal officer (LSO), said the T-45 will improve the student's ability to complete carrier qualifications. Unlike the TA-4J, which requires a fairly low maximum fuel for carrier landings, the T-45 can conduct carrier landings almost immediately after being catapulted with a maximum fuel load.

He estimates that the T-45 will burn less than 200 lb. during a normal carrier pattern and landing, compared to almost 500 lb. for the TA-4J. This capability would allow a student to make all the required landings without stopping for refueling.

At present, a student pilot makes four arrested landings in the T-2 and six in the TA-4J to carrier qualify. The T-45 is equipped with a catapult launch bar, and does not require the catapult bridle used on the TA-4J and found less and less frequently on Navy carriers.

During the flight debrief, Dubois also praised the Goshawk as an effective element of the T-45 training system. He said the T-45's flexibility will change the way the Navy trains pilots. He also said the T-45 is a much better solo aircraft than the other Navy trainers.

T-45 simulators offer major training benefits
David M. North/Kingsville, Texas

The ground-based portion of the U.S. Navy's T-45 training system encompasses realistic flight simulators, advanced electronic training devices and student scheduling and monitoring unique in military pilot training programs.

The simulators used in the Goshawk syllabus were developed by Hughes Training, Inc., under contract to McDonnell Douglas. The current Navy plan calls for a total of 16 operational flight trainers (OFT) and eight instrument trainers to be located between here and Naval Air Station Meridian, Miss.

Initially, the Navy had expected to receive 20 OFTs and 12 instrument trainers. The downward revised number of simulators is based on 450 Navy/Marine Corps pilots being trained annually.

Training Sqdn. VT-21's T-45s still show the USS Forrestal *on the side of the aircraft to denote the training carrier, which has been decommissioned.*

The Navy is now discussing numbers closer to 300 pilots receiving their wings annually, so the total number of required simulators could be reduced even further, The decrease in new-pilot demand is being driven by the fewer number of operational Navy air wings to be supported under a declining Defense Dept. budget.

The one offsetting factor against lowering the required number of simulators would be the surge need for pilots during a national emergency.

Hughes Training has delivered six of its OFTs and two instrument trainers to the Navy as of this summer. The instrument trainers do not have the visual system or the Smith Industries head-up display installed in the OFTs.

During the visit to the ground training facility here, I had the opportunity to evaluate one of the OFTs in operation. The OFT is equipped with a Hughes Simulation System SPX500HT computer image generation system with color and a partial dome screen. The OFT and instrument trainer use a Encore 32/8780 computer.

John Waples and Thomas McBrien aided me in flying the simulator. Waples is McDonnell Douglas' T-45TS business development manager at Naval Air Training Command, and McBrien is a Hughes staff engineer for the T-45 system.

The partial visual system in the OFT goes 30 deg. vertically up and down from the nose of the aircraft, and 105 deg. to the left and 75 deg. to the right of the aircraft centerline. The increased visual distance to the left allows for a normal carrier approach, air-to-ground weapon deliveries and formation flying. The trainers are equipped with an active g-suit motion cuing system and an aural system to provide aircraft motion and to help aerodynamic situational awareness.

I found that the Hughes OFT was realistic when flying two carrier approaches, one a touch-and-go and the other an arrested landing. All of the visual and aural cues were as I expected. Student pilots flying the TA-4J and the T-2 are given time in the T-45 OFT to acquaint themselves with the carrier environment. This program has been very successful, McBrien said.

Two weapon delivery patterns were flown delivering rockets in a 45-deg. dive to a simulated calibrated target. The aerodynamic performance of the T-45 during the roll-in and tracking of the target was extremely realistic. The firing of the rockets and the 60-ft. miss of the bull's-eye was gratifying to someone who has not done this in a long time.

The quality of the simulator and the associated training accompanying its use appeared to be professional. It would appear that once the Navy has brought a student to advanced training, it is not sparing expense or time to see that the pilot receive the best training available.

The simulator curriculum contains 72 flights for almost 100 hr. for each student. The flight missions range from cockpit orientation, night and day formation, operational navigation and air combat maneuvering to the final carrier qualification flights. Of the total, 51 flights are in the OFT and 21 in the instrument trainer.

Total flight time in the actual T-45 is scheduled to include 132 flights for 176 hr. Another 115 hr. are scheduled for briefings to support individual flight segments, and a total of 81 hr. are spent by the student in instructor-taught classrooms. A large portion of the 81 hr. is involved with student indoctrination and aircraft systems. The Navy estimates that a student's advanced training time is 203 working days, or 40.5 weeks.

The ground school portion of the syllabus is equally impressive in its professional approach to pilot training. Ground instructors use 100-in. rear projection screens in each classroom to display computer-generated views of the T-45 and its systems. Once aircraft familiarization and systems are taught, the instructor can show the view from the cockpit for every flight situation. This view can be analyzed and shown through the headup display. This concept is especially helpful when analyzing air combat maneuvering, air-to-air gunnery and ground weapon delivery.

The instructor portion of the ground training is augmented by individual computer-based learning centers. The trainees are able to replicate the same visuals used by the instructors on their own computer monitors.

The McDonnell Douglas support staff here does all of the upgrading to the course curriculum. The staff can revise both the software used to present the information or graphics and the academic material. Included on the McDonnell Douglas staff are a graphic artist and an audio-visual specialist. Much of the input for the requested changes are from instructors and students who see a better way of doing or portraying a certain flight segment.

To anyone in the military who has had to put together a flight schedule, the McDonnell Douglas training integration system (TIS) has a definite advantage over the manual method. My exposure to TIS started with logging a completed T-45 flight in the computer in the maintenance area. The maintenance data are logged for each aircraft, as well as flight time, landings and any other pertinent writeups.

The instructor's time is logged for his individual flight time record. In the case of a student, the flight completion is recorded, as well as the grades. This information is stored for the next day's flight schedule.

The instructor can input any restrictions for flying the next day. The TIS also schedules the next appropriate flight for the student, and assigns a qualified instructor. The TIS will not schedule a flight that does not meet the weather requirements. The integrated system also schedules classroom and simulator time. "A complete flight schedule for the next day takes 30 min. to generate," one McDonnell Douglas official said. "And that schedule includes all the possible variables you could imagine. The only variable we cannot include is whether the instructor and student get along well."

The TIS also includes performance records for all the students, and the system can be used for data analysis and reporting. This element is especially effective in tracking problems encountered by an individual student, or whether there is a failure trend developing in any one of the training segments.

The same system is used for administration management and virtually ties together all of the records of the training command here. The Navy's chief of naval air training in Corpus Christi and the chief of naval education and training have TIS terminals and access to the Kingsville data.

S.211A offers size, cost advantages

David M. North/Calverton, N.Y.
October 26, 1992

The SIAI Marchetti/Grumman S.211A's small size and light weight, overall performance and excellent pilot training capabilities make it a strong contender in the U.S. Joint Primary Aircraft Training System (JPATS) competition.

Grumman's and the Agusta SIAI Marchetti Div.'s offering for the U.S. Air Force/Navy's primary trainer role is similar to that of its competitors. The JPATS competition for 750-plus aircraft and potential orders from Canada and other countries have sparked one of the most widely contested military aircraft programs in recent years.

To meet the requirements for the U.S. military's primary trainer role, there have been an unprecedented number of international pairings. The majority of contenders are non-U.S. aircraft with U.S. partners providing the program management and varying levels of final assembly.

In addition to the Grumman offering, Lockheed has teamed with Italy's Aermacchi to offer a version of the MB339. Vought is working with Fabrica Militar de Aviones in Argentina to propose the Pampa 2000. These three jet-powered trainers have gained experience in service in the air forces of various countries. The S.211 is being flown by Singapore and the Philippines. The MB339 is flown by Italy and Australia, as well as others. The Pampa, which has a historical and technology tie to the Dornier Alphajet, is used for pilot training in Argentina.

Two turboprop aircraft also are being proposed for the JPATS program. Brazil's Embraer has teamed with Northrop to offer the Tucano H, and Beech Aircraft is in the process of building the Pilatus Aircraft PC-9 Mk.2 with a Pratt & Whitney Canada PT6A-68 engine as their entrant.

Both Air Force and Navy officials involved with the JPATS program strongly contend that the winning trainer will be picked on the aircraft's ability to meet mission requirements as well as the user's objectives, initial price and life cycle costs. "Whether the contending trainer is a turbofan or a turboprop has become basically irrelevant," one senior military official said. The ground-based training portion of JPATS also has been given a significant role in the selection process, with the aircraft teams signing on training companies (AW&ST Aug. 3, p. 36).

Two wild card aircraft possibilities are serious contenders for the JPATS award. Rockwell International, in a teaming agreement with Deutsche Aerospace, is building the Fan Ranger with the same Pratt & Whitney Canada JT15D-5C engine as is in the S.211A. The Fan Ranger is scheduled to first fly in December of this year or early next year.

The other wild card is being developed by Cessna Aircraft. The Cessna contender incorporates two Williams International FJ44 engines buried in the fuselage and a modified CitationJet aft fuselage. The business aircraft's wings have been clipped by some 10 ft., and a tandem cockpit replaces the cabin. First flight of the all-U.S. built aircraft is expected next summer.

Of all the primary jet trainers being offered for the JPATS program, the S.211A is the smallest but its gross weight is in the same general range as the Tucano H and the PC-9 Mk.2.

I first flew the basic S.211 from Sesto Calende, Italy, in 1985. That aircraft had straightforward handling characteristics, more than adequate performance and an ease of operation compatible with the primary trainer role (AW&ST Mar. 11, 1985, p. 66).

The primary difference between the S.211A flown from Grumman's Calverton, N.Y., facility and the S.211 flown in Italy is the substitution of the Pratt & Whitney Canada JT15D-5C for the earlier JT15D-4C engine. The S.211A offered for JPATS also will include greater fuel capacity and an electronic flight instrument system (EFIS). The –5C engine is rated for 3,190 lb. of thrust, some 690 lb. more than the 2,500 lb. of the –4C engine. Pratt & Whitney redesigned much of the –4C engine to achieve the additional thrust of the –5C and lowered the bypass ratio from 2.68 to 2.08. At the same time, to offset the lower bypass ratio, the cycle pressure ratio was increased to help counter the higher specific fuel consumption.

Empty weight of the final production aircraft also has increased from 4,330 lb. to 4,460 lb., while the maximum takeoff weight without external stores has been increased from 6,060 lb. to 6,393 lb. The S.211 flown in 1985 had a lower empty weight, but this was prior to the final configuration selected by the Singapore air force. The JT15D-5C engine accounts for approximately 100 lb. of the empty weight gain. The landing gear in the S.211A also has been modified to sustain a 13 ft./sec. sink rate.

The more powerful Pratt & Whitney engine was installed in the S.211A largely to improve field length requirement, time to climb figures, sustained g-load maneuvering above 20,000 ft. and to increase maximum speed by 50 kt., Philip Murphy, the Grumman manager of JPATS business development, said.

The S.211A was flown from Calverton with David Kratz, Grumman's project pilot for JPATS and an F-14 pilot in the Navy and Naval Reserve. Following a thorough preflight with Kratz, I occupied the front seat and Kratz the rear seat. The S.211A is equipped with a Martin Baker IT10LA ejection seat capable of a zero airspeed and zero altitude operation. Breakers on top of the seat penetrate the canopy during ejection.

The fuel on board the S.211A was 1,350 lb., near its maximum load. The Air Force will evaluate an S.211A with a fuel capacity of 1,500 lb. The gross weight of the trainer on the Grumman ramp was approximately 150 lb. less than its maximum of 6,393 lb.

Engine start was standard and after a slight delay while correcting a faulty radio connection, I taxied the S.211A to the active runway. I found the S.211A easy to taxi using the rudder pedals to control the nose wheel steering. Grumman and Agusta plan to install selectable nose wheel steering in the final S.211A JPATS candidate. The nose wheel activation switch is to be installed below the gun trigger on the forward portion of the control stick. The selectable steering is safer for primary students during high speed taxi and crosswind landings. The aircraft's brakes were effective and there was no tendency to grab at any time.

The temperature on the clear day was 66F, and the wind was from 330 deg. at 14 kt. I advanced the throttle to maximum power on Runway 32 while still holding the brakes without any slippage. The 1,500-ft. takeoff roll took 15 sec. Landing gear and flaps were raised in rapid succession, and the fuel remaining was 1,200 lb. while establishing a 200-kt. climb schedule at nearly a 200-ft. altitude.

I reached 5,000 ft. altitude 1 min. after takeoff from the sea level airport. Rate of climb passing through this altitude was 4,400 fpm, and the fuel flow was 1,550 lb./hr. I took 2.3 min. to reach 10,000 ft., and

SIAI Marchetti/Grumman S.211A is powered by a single Pratt & Whitney Canada JT15D-5C engine with 3,190 lb. of thrust.

the fuel flow was 1,300 lb./hr. while climbing at 3,750 fpm. The time to reach 15,000 ft. was 3.6 min. The fuel flow had decreased to 1,150 lb./hr. and the rate of climb to 3,400 fpm.

An altitude of 20,000 ft. was reached in 5.5 min. with the rate of climb registering 3,000 fpm and the fuel flow at 1,040 lb./hr. SIAI Marchetti data show that under similar conditions, the basic S.211 would take approximately 40 sec. longer to reach 20,000 ft. from sea level. The fuel required to reach 25,000 ft. from 200 ft. was 150 lb. Just prior to reaching 25,000 ft. 7.6 min. after takeoff, my rate of climb was slightly more than 2,200 fpm. Fuel burn was 950 lb./hr.

During the climb and after leveling off at 25,000 ft., I was able to acclimate myself to the S.211's cockpit. For a small aircraft, I found there was ample room in the front cockpit to reach down and manipulate controls on the side consoles. However, my earlier experience and largest amount of flight time was logged in cramped McDonnell Douglas A-4s, which might cloud my perception of cockpit roominess. The cockpit is more than adequate and is comparable in size to some of the other JPATS candidates I have flown.

I also prefer to sit high in the cockpit to improve outside visibility. In the S.211, sitting high in the seat brought the front canopy bow into my field of view. By lowering the seat, the canopy bow was less of a factor.

The pressurization of 3.5 psi. provided a 13,000 ft. cabin pressure at 25,000 ft. During the entire flight the air conditioning system was run in the automatic mode and provided a comfortable flow of air. SIAI Marchetti easily could increase the pressure differential to 5.0 psi. or more, Murphy said.

At 25,000 ft., a maximum range throttle setting of 85.6% N_2 was established, which gave a speed of Mach 0.45 and a fuel flow of 510 lb./hr. I then used a split-S maneuver to reach a lower altitude. The roll rate was quick entering the maneuver. Kratz said the maximum role rate of the S.211A is 160 deg./sec., very close to the 150 deg./sec. I achieved in the basic S.211. The maximum roll rate is achieved at 220 kt. and between 240 kt. and 250 kt.

At 15,000 ft. in the clean configuration, I slowed the S.211A to 110 kt. and with the gear and flaps down flew at 100 kt. In both configurations, pitch and roll control was positive with little sloppiness in the stick.

With the throttle at idle, the 100 kt. stall speed in a clean configuration was approached at a 1 kt./sec. deceleration. The S.211A stalled close to 100 kt. with no tendency for a wing drop. The stall was held until the aircraft had a nose high attitude and an increasing rate of descent. Much the some happened in the dirty configuration with an 85-kt. stall speed. There was virtually no aerodynamic warning prior to reaching stall conditions.

Grumman is testing an off-the-shelf stall warning system made by Globe Motors. The Globe stickshaker could be set to a predetermined angle of attack to give the student a stall margin, Murphy said.

After returning to 15,000 ft., power was reduced to idle. At 105 kt. in the clean configuration and with almost a 20-deg. pitch attitude, the stick was brought full aft and full right rudder was introduced. The S.211A initially turned left, hesitated and then went into post stall oscillations to the right.

The maneuver could not be labeled a stabilized upright spin, because the turn needle was not pegged, the airspeed was fluctuating and the pitch changed. After two turns, the S.211A was in a more stabilized spiral with the airspeed increasing. A full three turns were made before the stick was neutralized with opposite rudder input and the aircraft was pulled out in approximately one half turn with a total loss of 2,100 ft. Kratz said the S.211 would recover easily with hands off in one turn. The speed buildup in the spiral would make it difficult to hold in pro-spin input for long. Grumman officials believe the spin/spiral characteristics of the S.211A are not a detriment in primary training, but could add aerodynamic fixes if rigid upright spin characteristics are mandated.

A loop was started at 10,000 ft. at 250 kt. and 85% N_2 power rating. An initial 3.5g pull resulted in a 3,000-ft. gain in altitude over the top

Student occupies the front 2.211A cockpit.

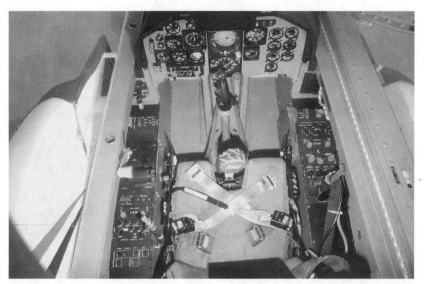

High position of the rear cockpit seat allows good visibility for the instructor.

of the loop. In this and other aerobatic maneuvers, I found the aircraft to be very clean and crisp in handling, with the outcome of the maneuver predictable, even with my lack of recent aerobatic experience.

I slowed and stabilized the S.211A at 200 kt. and 10,000 ft. to perform an acceleration check. With maximum power, it required 23 sec. to reach 250 kt. and 56 sec. to attain 300 kt. The maximum speed of 323 kt. at that altitude was reached in 135 sec. While at the same altitude and full power, the S.211A was banked 60 deg. and pulled to sustain 4.59 while stabilizing at 250 kt. to check sustained g-load ability.

At 15,000 ft. I was able to sustain 3.5g while maintaining 205 kt. in the turn. The S.211A has been modified to sustain +7g and −3g. This is an increase of +1g from the basic S.211.

SIAI Marchetti has kept the flight controls of the S.211A as simple as possible, using conventional push-pull rods connected to the dual sticks and rudder. The below-the-fuselage speed brake is electrically controlled and hydraulically actuated. The speed brake can be positioned to intermediate positions. Actuation of the speed brake at 250 kt. did give a slight ballooning to the S.211A.

The speed brake was again used during a 45-deg. dive and just prior to reaching the maximum 400-kt. speed. After correcting with nose down pitch, the speed stabilized at 400 kt.

The instrument panel of the S.211A I flew has standard round dials with an effective layout for scan. As with the other competitors,

Grumman and SIAI Marchetti will offer an electronic flight instrument system. The EFIS is being installed in the second production S.211A prototype in Italy for delivery to the U.S. early next year. The first production prototype flew for the first time on Sept. 17 in Italy and will be used as the test aircraft.

I flew the S.211A back to Calverton from the evaluation area over the Atlantic Ocean south of Long Island. A Navy-type break over Runway 32 was flown, using the speed brake to slow to 120 kt. at the abeam position. The trainer was very comfortable in the landing pattern, with excellent visibility from the front cockpit. Speed control was precise without requiring constant attention.

In the short straight-away position, the speed was 110 kt. with full flaps. The flare was initiated at 100 kt. with the touchdown speed of 95 kt. Two touch-and-goes were performed with the same smooth landings.

Kratz suggested that the final landing be a simulated engine out approach. The high key point of 3,000 ft. above the runway and on runway heading was reached at 130 kt. with the power at 55% N_2 and the speed brake extended. I made the downwind turn about 3,000 ft. down the runway. The abeam position was at 1,400 ft., still at 130 kt. While turning to the final runway heading at the 90-deg. position, I realized that I had flown too wide in the abeam position and could not safely land without adding power. Kratz said that with a little practice, the flame-out approach was an easy maneuver. The landing again was smooth, even with the abnormal approach.

The S.211A was rolled without braking to the end of the runway to be near the Grumman ramp. Kratz said that a 1,500-ft. landing roll was easily attainable with moderate braking. Total flight time was 1.2 hr., and the blocks-to-blocks time was 1.5 hr. Total fuel consumption was 950 lb.

The S.211A's ease of handling and overall performance characteristics makes it an ideal candidate for the primary training role. Whether the aircraft's spin characteristics have to be changed will be up to the services.

The S.211A's small size and lower end of the gross weight spectrum should allow Grumman to be competitive in acquisition cost. The simple design philosophy, plus the reliability of the Pratt & Whitney Canada JT15D engine should help the U.S./Italian team offer relatively low life-cycle costs.

The uncertainty surrounding the restructuring of Italy's aerospace industry is not expected to have a negative impact on the SIAI Marchetti/Grumman program. Some Grumman officials anticipate that SIAI Marchetti might become a division of Alenia, a producer of fixed-wing aircraft. The terms of the contract signed in 1988 call for a 50-50 work split, with Grumman building the front fuselage of the S.211A and doing final assembly, and SIAI Marchetti producing the wing, aft fuselage and tail.

SIAI MARCHETTI/GRUMMAN S.211A SPECIFICATIONS

POWERPLANT
The S.211A is powered by one Pratt & Whitney Canada JT15D-5C turbofan engine producing 3,190 lb. of thrust on a standard day.

WEIGHTS
Empty weight	4,453 lb. (2,020 kg.)
Maximum takeoff weight/clean	6,393 lb. (2,900 kg.)
Maximum landing weight	6,393 lb. (2,900 kg.)
Maximum takeoff weight/ext. stores	7,716 lb. (3,500 kg.)

DIMENSIONS
Wing span	27.66 ft. (8.43 meters)
Length	31.16 ft. (9.50 meters)
Height	13.45 ft. (4.10 meters)

PERFORMANCE
Maximum airspeed	400 kt. or Mach 0.8
Landing gear refraction/extension	160 kt.
Speed brake extension	unlimited
Maneuvering G-forces	+7, −3
Maximum crosswind landing component	25 kt.
Takeoff distance, 6,393 lb., sea level, ISA	1,450 ft.
Landing ground run, 5,644 lb., sea level, ISA	1,400 ft

Powerful Tucano H
handles well at high, low altitudes

David M. North/Sao Jose Dos Campos, Sao Paulo, Brazil
April 20, 1992

The Tucano H proof-of-concept aircraft offers preview of the performance available in the Embraer candidate for the Air Force/Navy Joint Primary Aircraft Training System (JPATS) competition.

The Brazilian aircraft manufacturer started development of a modified Tucano in February, 1991, after realizing that its standard Tucano would not meet the stringent performance requirements for the U.S. services' next primary trainer. The proof-of-concept Tucano H that first flew in September, 1991, is the same aircraft that this *Aviation Week & Space Technology* pilot flew from the Embraer headquarters here.

The first hurdle Embraer faced in its quest for increased performance was the installation of a new turboprop engine in a standard Tucano fuselage. The Tucano A is powered by a Pratt & Whitney Canada PT6A-25C engine with a 750-shp. rating. The proof-of-concept (POC) Tucano H has a 1,600-shp. engine installed, and the prototype Tucano H will have a PT6A-68 engine, with the same power rating and a similar installation.

More than doubling the power rating in the Tucano required a 4.5-ft. (1.37-meter) fuselage extension to restore the trainer's center of

Embraer Tucano H proof-of-concept trainer is powered by a 1,600-shp Pratt & Whitney Canada engine with a five-blade Hartzell propeller. The Tucano A fuselage was lengthened by 4.5 ft. to accommodate the higher thrust engine.

gravity and stability. A 3.3-ft. (1-meter) section was added to the aft fuselage in front of the empennage, and 1.2 ft. (370 cm.) was added in front of the engine.

Cooperation with Northrop

"Once we proved that the POC aircraft could handle the 1,600-shp. engine, the rest of the requirements for JPATS are more straightforward," Horacio Aragones Forjaz, Embraer's senior vice president of engineering, said. "The available power will easily offset the weight added for pressurization, and the required glass cockpit is an off-the-shelf system."

Embraer does not have any plans at this time to reduce the power rating of the PT6A-68 engine. The company also plans to exceed the minimum 3.5 psi. requirement for pressurization for the primary trainer candidate.

Once the POC flight test proved the Tucano H a viable contender for the JPATS program, Embraer started to look for a U.S. partner. A number of U.S. companies were approached, including General Dynamics and Northrop. As a reflection of the decision, a number of Northrop employees already were working on the program here.

"The choice of Northrop as a partner was the right one from a technical and practical point of view," Ozires Silva, president of Embraer, said. "We had built tail assemblies for the F-5E some six years ago, and we developed a good working relationship with them. We evaluated all the political, technical and management issues and feel we can continue to have a good working relationship with Northrop."

Embraer signed a 90-day letter of intent with Northrop to evaluate the Tucano H as a contender for JPATS. The 90-day letter will expire in May. If Embraer and Northrop reach agreement by then, the Brazilian company will build two prototype Tucano Hs and ferry them to the U.S. later this year. The two prototypes will have PT6A-68 engines installed and much of the other JPATS-required equipment.

If the Tucano H wins the JPATS competition, much of the aircraft would be completed in the U.S. Sections of the fuselage and wing assemblies would be built in Brazil and shipped to the U.S. Northrop would install the engine, systems and electronics and perform the final assembly. The U.S. company also would be responsible for flight testing the trainer. The aircraft would not necessarily be assembled at Northrop's facility in California, as sites in Florida also are being evaluated, Ozires Silva said.

While the Tucano H proof-of-concept trainer is not the final version that will compete for JPATS, its performance closely matches that of the prototype aircraft. The POC, however, has some limitations that will be eliminated in the prototype aircraft.

These restrictions were explained to me by Embraer's chief test pilot, Gilberto Pedrosa Schittini, before the evaluation flight. The primary restriction was that I was not to fly inverted or experience negative g-forces for any length of time. The engine gear box in the PT6A-67R has not been cleared for inverted flight. In two instances, the gear box has been damaged during negative g-force flight and the engine had to be shipped back to Pratt & Whitney Canada. An alarm has been installed to warn of an impending problem.

Following the preflight briefing and a walkaround, I strapped into the front seat. Schittini took the rear seat. The POC is equipped with a zero-zero Martin Baker Mk.10 ejection seat. The prototypes will be equipped with the same seat or one with similar performance. The Tucano A has a Mk.8 ejection seat that does not have the same performance.

The POC weighed 6,390 lb. (2,900 kg.), some 1,120 lb. (510 kg.) below the maximum utility takeoff weight. The maximum takeoff weight of the Tucano H prototype will be higher, but the actual figure has not been determined. The fuel on board was 1,140 lb. The instrument panel was standard Tucano, but an electronic flight instrumentation system will be added to the prototypes.

Engine start was normal with the use of a power cart. The throttle was brought forward at 15% power, and engine temperature was well within limits during start.

The power of the 1,600-shp. engine was immediately noticeable during the initial taxi. With small advances of the single-lever throttle, the aircraft started to jump ahead. This was true in flight as well. On the ground, Embraer has installed a taxi switch that reduces propeller pitch so there is less need to use the brakes or reverse thrust. Nose wheel steering was effective through the use of the rudder pedals, and there was no tendency to oversteer. A nose wheel control handle will be added to the prototypes. The brakes were effective. There was no tendency to grab during turns.

In the POC, Embraer has installed an on-board oxygen generating system (OBOGS), and a similar system or liquid system will be put in the Tucano H. The trainer also was equipped with an air cycle conditioning system, replacing the Freon system in the standard Tucano. A small air intake is mounted on top of the fuselage behind the cockpit for the air cycle machine. It took a short time in the air before the cockpit cooled on this relatively warm day.

The pretakeoff check list was minimal for the POC, but did require putting the propeller pitch switch in the flight mode.

I lined the aircraft up on Runway 15 at the Sao Jose dos Campos airport. Although it is a commercial field, there are only a few airline flights daily. The temperature was 24C (75F), and the wind was from 270 deg. at 2 kt.

The brakes were held until reaching 50% torque as the throttle was advanced to takeoff power. The POC leaped forward at brake release and acceleration was quick. There was an initial tendency to drift left as I overcompensated for the effect of the torque. Two units of right rudder already had been set into the rudder trim. Takeoff roll was about 800 ft., and when the landing gear and flaps had been raised, the POC already was at its climb speed of 160 kt.

An altitude of 5,000 ft. msl. was reached in less than 1 min. from the 2,000 ft. airport altitude. It required 1.5 min. to reach 10,000 ft. from 5,000 ft., or slightly less than 2.5 min. from brake release. The rate of climb at 10,000 ft. was 3,500 fpm and the fuel flow was 650 lb./hr.

The rate of climb at 15,000 ft. was 2,500 fpm. Slightly less than 3.5 min. was needed to reach that altitude from 5,000 ft. Fuel flow climbing through 15,000 ft. was 560 lb./hr. with the same 160 kt. climb speed. Shortly after passing through 15,000 ft., a 180-deg. turn was required to stay in the operating area and miss a thunderstorm that was quickly building during the late afternoon flight. These maneuvers lengthened the climb time to 20,000 ft.

A total of 8 min. had elapsed from takeoff to reaching 20,000 ft., and the fuel burned from the ramp to altitude was 140 lb. These climb results are far better than that of the standard Tucano A I flew in Brazil in 1983 (AW&ST Sept. 19, 1983, p. 51). During that flight, the rate of climb at 122 kt. at 5,000 ft. was 2,100 fpm and at 10,000 ft. was 1,600 fpm. It also took 4 min. to reach 10,000 ft., 9 min. to 15,000 ft. and 17 min. to 20,000 ft.

Performance of the Tucano T Mk.1, which is flown by the Royal Air Force, is on the high side between the Tucano A and the Tucano H proof-of-concept trainer. The RAF Tucano is powered by an 1,100-shp. Garrett TPE331-12B turboprop engine. There are other differences as well between the Brazilian Tucano and the one Short Brothers builds for the RAF.

The level flight performance of the Tucano H also reflected that it has more than twice the power of the standard Tucano. The outside air temperature at 20,000 ft. was more than 20C (68F) warmer than standard. At a full 60% of torque, the indicated speed of the trainer was 208 kt., and the fuel flow was 500 lb./hr. The gross weight of the POC was 6,250 lb.

Pulling the throttle back to a 40% torque setting to approximate a long-range cruise speed resulted in a 191 kt. indicated speed and a fuel flow of 390 lb./hr. A split-S maneuver, taking care not to remain inverted long without positive g-forces, was used to descend to 15,000 ft. At this altitude, I made hard turns in both directions and found the trainer to be responsive, with no buffet at a 60-deg. bank and pulling 4g. Several aileron rolls also were made with the nose at 5-deg. nose up and 200 kt. The roll rate was sharp for a turboprop trainer, but there was some buffet during the coupled rolls. Schittini said that in the POC Embraer had increased the aileron travel to 25-deg. from 18-deg. in the Tucano A. This increased the roll rate, but the leading edge of the aileron protruded below the wing and caused the buffet. He said that the initial goal was a 180 deg./sec. roll rate, but the aileron travel would be reduced to achieve a 150 deg./sec. roll rate.

During the maneuvers, I found that the visibility from the front cockpit, as it was in the Tucano A, was excellent. When Embraer installs a new canopy to achieve JPATS bird strike requirements, the height of the canopy will be increased to allow both the seats to be higher. This will provide for increased over-the-nose visibility.

A unique performance feature that I found to be distracting during flight was that a small amount of movement on the throttle resulted in larger-than-expected speed changes and required compensation with rudder. Schittini said that the prototype Tucano H will have a FADEC (full-authority digital engine control) for the PT6A-68 to smooth out engine response. There will be a mechanical back-up fuel control. The prototype also will be equipped with an automatic rudder trim system to lessen pilot workload.

The POC is equipped with a five-bladed Hartzell propeller. Schittini said that because of the gyroscopic mass of the five-blades, it is possible the Tucano H will have a four-bladed propeller.

Power was reduced to near idle to enter a stall in the clean configuration. While slowing at a rate of 1-kt./sec., the trainer stalled at 75 kt. In the landing configuration the stall was achieved at 68 kt. In both cases, there was no tendency to fall off on a wing, and the aircraft was controllable with an ensuing higher rate of descent. Stall speed of the Tucano A was slower by about 5 kt. in my 1983 flight.

A spin was entered by reaching near stall speed with the stick full aft and with an input of rudder. Recovery was accomplished from the rather benign spin after two turns by centering the stick and applying opposite rudder. Schittini said that spins in the POC tended to be more vertical than in the standard Tucano because of the higher weight in the nose. While at 13,000 ft. and full power, a 4g loop was entered at 200 kt. Speed at the top was 100 kt. at 16,000 ft.

The speed brake in the POC was used during the descent to a lower altitude. At the higher speeds, the speed brake on the lower fuselage did cause a slight ballooning when deployed.

With full power at 300 ft. above the Brazilian terrain, the POC's speed reached 275 kt. on the warm day. The Tucano H was responsive to pitch and roll controls at this altitude, and the trainer felt like a heavier aircraft in light turbulence.

A touch-and-go was performed at the airport, followed by a full-stop landing. Both patterns were flown close to the runway, with a 120 kt. abeam speed and 100 kt. over the runway. The responsiveness of the aircraft's controls and engine plus good visibility over the nose allowed for two smooth landings, although I had to search for the runway on the first landing.

Total flight time was 55 min., and blocks-to-blocks time was 1.1 hr. The fuel used during the flight was 417 lb.

I have flown the majority of the candidates for the JPATS program, although not all of them have been in their final configuration. Both the Tucano H and the Beech/Pilatus Aircraft PC-9 turboprops offer near jet-like performance with single throttle lever control. The Air Force is historically less apt to choose a turboprop trainer than the U.S. Navy. However, the performance of the turboprop trainers, a potentially lower initial acquisition cost and lower operating costs are likely to be strong factors in their favor. This is especially true in the current budget climate.

Pilatus upgrades trainer performance systems with PC-9 turboprop

David M. North/Stans, Switzerland
July 22, 1985

Impressive flight handling qualities, increased performance and new aircraft systems compared with its predecessor mark the Pilatus Aircraft, Ltd., PC-9, the second turboprop trainer offered by the Swiss manufacturer for the world basic through advanced pilot training market.

Both the No. 1 and No. 2 prototype PC-9 trainers were flown from the Pilatus manufacturing facility here by this *Aviation Week & Space Technology* pilot. The two flights with Hans Galli, chief test pilot for Pilatus, were to evaluate the PC-9's handling characteristics and assess the differences between the PC-7 and the newer PC-9. The PC-7 had been evaluated by *Aviation Week & Space Technology* in 1983 (AW&ST Dec. 19, 1983, p. 65).

Although the PC-9 and the PC-7 have a similar configuration, the newer aircraft is powered by a Pratt & Whitney Canada PT6A-62 turboprop engine derated from 1,150 shp. to 950 shp. The PC-7, introduced in 1976, has a 550-shp. PT6A-25A. The remainder of the airframe and systems of the PC-9 have approximately a 15% commonality with the earlier trainer.

The black painted No. 1 PC-9 trainer was flown initially. Primary differences between the two prototypes involve the systems and instrumentation in each aircraft. The No. 1 prototype, which first flew in May, 1984, is equipped with a head-up display and standard flight instruments, while the second prototype has an air-conditioning unit, nose wheel steering and electronic flight instrument system (EFIS) displays.

Ramp weight of the No. 1 prototype on the warmer than standard day was close to its 4,982 lb. maximum for aerobatic flying. An additional 2,000 lb. can be carried by the trainer in external stores and fuel tanks for the utility mission. The aircraft carried a full fuel load of 950 lb.

Following the preflight of the PC-9 on the ramp, I strapped into the front seat and Galli took the rear seat. Unlike the PC-7, the newer trainer is equipped with two Martin-Baker MK. CH 11A ejection seats with integrated personal survival packs. The lightweight ejection seat has an operating envelope of 60-400 kt. The minimum speed was selected, Galli said because of the 80-kt. rotation speed of the aircraft

Pilatus PC-9 is equipped with Martin-Baker ejection seats and powered by a single Pratt & Whitney PT6A-62 turboprop engine rated at 1,150 shp and derated to 950 shp.

and because the aircraft is stressed so that a pilot could survive a crash at 60 kt.

The single PT6A-62 engine was started by use of the aircraft's battery, and the power control lever was moved from off to idle position when the engine gas generator speed topped 10%. The PC-9 is equipped with a single control for the operation of the fuel schedule and propeller speed.

The propeller speed is set at 2,000 rpm at normal engine power settings. A constant speed unit adjusts the propeller pitch to match changes in flight conditions and power settings.

Fuel flow of the engine on the Pilatus ramp was 110 lb./hr. Nose-wheel steering by use of the rudder pedals was not incorporated in the No. 1 PC-9 prototype but was installed in the second trainer and will be installed on production aircraft. I found the ground control of the aircraft by differential braking to be more than adequate during the 7-min. taxi to the operational runway. A total of 20 lb. of fuel was used during the 9 min. from engine start to takeoff roll.

Acceleration during the takeoff roll was rapid, and the trainer used close to 800 ft. of runway to reach the 80-kt. rotation speed. The distance to clear a 50-ft. obstacle was estimated to be 1,500 ft. The landing gear was raised immediately after establishing a positive rate of climb with a 10-deg. nose-up attitude. The flaps were raised at 100 kt., and a 140-kt. climb speed was achieved a short time later.

We reached 7,000 ft. in less than 2 min. from starting the takeoff roll at the 1,500 ft-msl. airport. The initial rate of climb was 4,000 fpm, and the fuel flow was 520 lb./hr. The rate of climb decreased to 3,200 fpm at the 10,000-ft. mark with a 500 lb./hr. fuel flow a minute later.

Another 1.5-min. was required to reach 15,000 ft., where the rate of climb registered 2,600 fpm. The fuel flow was now 460 lb./hr. Fuel flow dropped to 400 lb./hr. at 20,000 ft. after another 2 min. of flying. The rate of climb at 20,000 ft. was still slightly in excess of 2,000 fpm. During the climb to a cruising altitude of 25,000 ft., clearing turns were made to observe traffic. The visibility from the cockpit of the PC-9 was excellent in observing the mountain scenery below and the airline traffic above the trainer.

The cruising altitude of 25,000 ft. was reached in slightly less than 9 min. from the beginning of the takeoff roll. We held a speed of 140 kt. during the climb to the aircraft's maximum operating altitude. A total of 90 lb. of fuel had been used from engine start, or 70 lb. from takeoff. The higher power of the PT6A-62 engine in the PC-9 gives it almost twice the rate of climb of the PC-7 and time-to-climb figures almost half that of the older trainer.

Maximum cruise speed

Maximum climb and cruise power were retained at 25,000 ft. until reaching a maximum cruise speed of 204 kt. indicated. The calculated true airspeed for the near standard day at altitude was in excess of 300 kt. Fuel flow at maximum cruise speed was 320 lb./hr., and the aircraft was at a 4,870 lb. gross weight.

An economy fuel flow setting of 230 lb./hr. resulted in a 176-kt. indicated airspeed and a true airspeed of approximately 250 kt. Galli said the maximum cruise speed was the normal speed used by the Pilatus pilots for longer-distance trips. At 25,000 ft., the range of the PC-9 is nearly 1,000 naut. mi. With two underwing fuel tanks, the trainer's range increases to nearly 2,000 naut. mi.

During the takeoff and climb, the engine temperatures and torque readings were monitored to see that the higher limits were not exceeded. The engine control includes an electronic unit that automatically limits these operating parameters in the maximum power settings. In case of failure of the automatic fuel metering system, an override system allows manual control of the valve metering fuel to the engine.

Control surface operation

While still at altitude, I executed a number of steep turns followed by wingovers. The primary control surfaces of the PC-9—aileron, rudder and elevator—are all manually operated from the front and rear cockpits by the use of cables, rods and bellcranks. Pilatus has achieved an excellent harmony between control input and expected response from the trainer in pitch and roll. The PC-9 is equipped with an electric trim system for the primary flight controls.

Aileron and four point rolls were accomplished next. The trainer, with maximum 20-deg. up aileron deflection and 11 deg. down deflection, has a roll rate of approximately 90 deg./see. at 170 kt. Both pitch and roll responses during these maneuvers were positive and quick throughout a wide range of speeds. The trainer incorporates a fuel tank that allows for inverted flight during negative g-load conditions for at least 60 sec. A loop started at 230 kt. at 17,000 ft. with maximum climb power, and using close to 4g during the pullup topped at 19,600 ft. and 190 kt.

The optional Jet Electronics & Technology, Inc., Series 1000 head-up display was used to monitor heading and wing position throughout the loop and was found to be easy to follow. Galli said there were still some symbology changes to be incorporated into the display, but pilots flying the system have found it easy to use, especially during landings.

The hydraulically operated airbrake, mounted on the underside of the trainer's fuselage, was used to slow the aircraft prior to performing stalls. The airbrake extends to 70-deg. deflection in less than 1.5 sec. and provides a pitch-up movement to the nose. Galli said the shape of the airbrake was to be changed so that the center of pressure would be further forward, eliminating the pitch-up but retaining the effectiveness of the airbrake. The PC-9's airbrake, not installed on the PC-7, halves the time required to slow from 300 kt. to 150 kt., Galli said.

Higher wing loading of the PC-9 and a slightly different wing design from the PC-7 has increased the stall speed of the newer aircraft.

At 15,000 ft. and a power setting to achieve a deceleration of 1 kt./sec., the PC-9 stalled at 78 kt. in the clean configuration. The stall was preceded by a very slight buffet and an intermittent high-tone stall warning signal. There was aileron and pitch control throughout the stall and recovery. The aircraft exhibited no tendency to either pitch over or drop off on a wing during the sequence.

The stall speed of the PC-9 with the flaps in the landing position and landing gear extended occurred at 62 kt., almost 8 kt. less than predicted by Galli. The angle of attack indicator in the front cockpit read less than 15-deg. at the time of the stall. The stall sequence in this configuration was the same as in the clean configuration, with a loss of altitude of less than 100 ft. after power was increased.

A maximum power-on stall at 10,000 ft. provided a much more unconventional stall sequence. On Galli's instructions, I pulled the nose of the PC-9 up to close to a 50-deg. nose-high attitude after starting from a relatively slow speed. The close to one-to-one thrust to weight ratio of the PC-9 allowed us to climb several thousand feet prior to registering a zero airspeed. The aircraft then rolled to the left and started to pitch down to what became a 70-deg. dive. The use of right aileron as the aircraft was accelerating allowed me to bring the wings level as the nose of the aircraft passed through the horizon. Approximately 400 ft. of altitude was lost during the maneuver.

The power then was reduced to enter a spin. At 80 kt. the stick was pulled straight back and full left rudder was applied. The first rotation was nose high, gradually turning into a stable spin with a 45-deg. nose-down attitude after four turns. The time to complete one turn was 3.2 sec. Galli said that a hands-off spin recovery could be done in two turns. Opposite spin rudder and placement of the stick in a slightly forward of neutral position provided for a recovery from the spin in one turn and with a loss of less than 5,000 ft.

Following recovery from the spin, I applied maximum power to the PC-9 at 11,500 ft. at 130 kt. It took 1.5 min. to reach 250 kt. The increased power of the Pratt & Whitney engine in the PC-9 was evident in this type of application, as well as in the time to climb and maximum speeds in comparison with the PC-7, Galli said. Pilatus guarantees the PC-9 for a maximum speed of 270 kt. at sea level and 300 kt. at 18,000 ft.

At 15,000 ft., the PC-9 was pulled into a 3g turn at 170 kt. without the speed decaying at maximum cruise power. At sea level, the PC-9 is capable of performing sustained 4g turns while flying between 170 and 200 kt. Galli said the PC-9 is not as speed dependent as the PC-7 while performing maneuvers with high g-loads.

Cockpit instrumentation

During the flight back to the Pilatus facility at Stans, I was able to observe the cockpit instrumentation. The general layout is similar to the PC-7's, but there are many significant differences. Standard round dial engine and secondary instruments in the PC-7 and the two PC-9 prototypes are to be replaced with a panel of liquid crystal displays on the right side of the instrument panel. Digital and analog displays will provide torque, internal turbine temperature and engine gas generator speeds. A combination of digital and analog displays will be used for fuel quantities, oil temperature, air inlet temperature and other indications.

The cockpit has been designed to facilitate the transition to a high-performance military aircraft. The single power control, the placement of the instruments and the options of a head-up display and electronic flight instrument system displays for the trainer reflect the efforts Pilatus has made to ease this transition for military pilots. The Swiss company also has allowed space on the lower left portion of the instrument panel and the side consoles for optional equipment so that operators can tailor the cockpit to their requirements.

VHF/UHF integration

Another feature I found to be practical was the integration of the King VHF and UHF radios in the aircraft. Galli conducted all of the radio transmissions during the flight from the rear cockpit. When a radio frequency was changed from the rear seat, it also was changed in the front seat radios, with the frequency shown in the digital display. Both cockpit radios would always be on the same frequencies, eliminating any confusion in the event of an intercom system failure.

The fuel flow at 15,000 ft. was 220 lb./hr. while indicating 160 kt. The noise level in the cockpit during cruise and at higher speeds during aerobatic maneuvers was comfortable even after removing the helmet for a short duration. The oxygen system in both cockpits is of the diluter/demand type and has nearly 50 cu. ft. of capacity.

Galli took control of the aircraft to perform one maneuver over the airport boundary at Stans. Starting at 280 kt. he pulled the PC-9 into a vertical climb and performed three rolls prior to reaching 120 kt. At the top of the climb he pulled the aircraft inverted and flew it out of a slight dive.

Airport pattern

Visibility from the cockpit for the training mission was excellent throughout the flight. This and the overall handling characteristics of the aircraft became important as I descended to the airport landing

Pilatus PC-9 cockpit is equipped with Collins electronic flight instrument system (EFIS) displays as an option. The Swiss company also offered a head-up display as an option on the turboprop trainer. A single-lever power control is used in the trainer.

pattern. The downwind to the active runway at Stans required a close-in pattern to avoid the small mountain that parallels the runway and rises almost vertically 2,000 ft. above the airport. The right wing tip was approximately 50 ft. from the trees on the mountain.

The relatively tight pattern required for the landing was not uncomfortable in the PC-9. Landing gear and flaps, both hydraulically powered, were lowered at the abeam position with landing flaps selected on the short straightaway. The final was flown at 95 kt., and I made a relatively smooth first landing. The PC-9 was easily turned on the runway to backtrack to the taxiway. Galli said there were few instances where reverse thrust could be used in the trainer, and the ability to keep the propeller control in the aircraft simple outweighed any advantage gained by reverse thrust. The landing ground roll was less than 1,000 ft. using nominal braking.

Total fuel burn during the 1 hr. 15 min. from blocks to blocks was 353 lb. The actual flight time was 1 hr. 3 min. Following the flight, I sat in the rear seat to observe the visibility afforded the instructor pilot. I was able to see most of the instruments on the left and right side of the front panel. All but a small portion of the nose of the aircraft was visible, allowing for a good perspective for the instructor demonstrating a landing from the rear seat.

The rear cockpit is almost identical to the front cockpit. The one primary difference is that the control for the landing gear emergency extension system is located only in the front cockpit.

I strapped into the front cockpit of the yellow and red No. 2 prototype for a short flight to evaluate the nose wheel steering and air-conditioning systems that will be installed in production aircraft. The second PC-9 also had Collins EFIS attitude and horizontal situation indicators. The EFIS displays will be optional on the trainers. Galli said that if an operator chose a J.E.T. head-up display for the PC-9, the attitude information could be portrayed on the head up display and Pilatus would recommend the installation of only one EFIS display. The single Collins unit would be capable of displaying, both attitude and horizontal information.

The nosewheel steering, controlled by a selector switch on the stick, was found to be sensitive at first with a tendency to overcontrol during the taxi to the active runway. Control of the nosewheel steering on the return to the Pilatus ramp was smoother. The second takeoff was much the same as the first, although this time I stayed in the landing pattern at the airport. The standard air cycle and two-stage heat exchanger were very effective in keeping the cockpit cooled during the flight at low altitude and above standard day conditions.

Full-stop landing

Two touch-and-go landings and a fullstop landing were accomplished in a 15 min. flight. The Collins EFIS displays were found to be readable and bright, but during most of the short flight I found that my attention was outside the cockpit and that the runway and angle of attack indicator above the glare shield were used as references for the landings. All three landings were smooth, with the final one the best of the four made during the two flights.

A total of 112 lb. of fuel was used during the 29 min. from blocks to blocks. During the two flights in the PC-9, I found the trainer to be agile and to have a takeoff, climb to altitude, top speed and maneuvering performance greater than other turboprop trainers I have flown. I have not flown the Pilatus PC-7, but the company's own performance charts indicate that the PC-9's performance is far better than the earlier aircraft (AW&ST Dec. 19, 1983, p. 65). At the same time, the performance of the PC-9 is higher than that of the Embraer EMB-312 Tucano, as one would expect of an aircraft having an engine with an additional 200 shp. (AW&ST Sept. 19, 1983, p. 51). The PT6A-25A engine powering the PC-7 is rated at 550 shp., and the PT6A-25C engine in the Tucano has a rating of 750 shp.

Engine selection

The Garrett TPE331-12B turboprop engine selected by Britain's Royal Air Force to power the Tucano trainer to be built by Embraer and Shorts Brothers will have a rating of 1,100 shp. The Garrett-powered Tucano, once flying, and the Pilatus PC-9 will be in the same performance class of trainers.

Pilatus had been at least one year into the development of the PC-9 when the Royal Air Force came out with a requirement for a primary trainer, according to Karl E. Scheuber, sales manager for the U.S., Canada, Central America and East Europe.

"Pilatus had started development of the PC-9 as a logical step beyond the PC-7," Scheuber said. "We had committed the PC-9 to production prior to the Royal Air Force stating its need. The British then came to us and looked at our performance figures, and based their requirements closely to what we were saying our trainer would do. The British government then picked the Embraer Tucano, but with a bigger engine to equal the PC-9's performance."

Joint production

British Aerospace had teamed with Pilatus us for the joint, production of the PC-9 for the Royal Air Force proposal. Pilots from the British aircraft manufacturer are still flying the PC-9 prototypes to define the performance of the new trainer and determine whether British Aerospace will Cooperate with Pilatus on any further sales efforts.

The Swiss aircraft manufacturer started looking at a single turboprop trainer in the 1960s but was unable to identify a strong market for the aircraft. Pilatus initiated the development of the PC-7 in the early 1970s, flying the trainer in 1976 for the first time and beginning deliveries in 1978. Since beginning PC-7 deliveries, Pilatus has gained a majority of the international turboprop trainer market by delivering nearly 350 aircraft to more than 15 operators.

Included in the customer list for the PC-7 are the air forces of Austria, Chile, Bolivia, Guatemala, Iran, Switzerland and Burma. Mexico operates 58 of the turboprop trainers, Iraq close to 50, Malaysia 44 and the Swiss Air Force 40.

The PC-7 received civil certification in 1983, and two of the trainers are flying in the U.S., one in Switzerland and two in France with civil registration. During my trip to Stans, the second PC-7 was being delivered to a U.S. customer. Scheuber said that Charles Nogle of Champaign, Ill., the president of the Beech T-34 club, had just finished flight training and was taking delivery of his turboprop trainer.

Additional orders

Scheuber said that almost every customer for the PC-7 has returned to order additional aircraft for its air force. Pilatus reached a high yearly production rate of 84 aircraft in 1983, dropping to 40 this year. The company estimates that it will build 40 PC-7s in 1986 and 1987. The difference in price between the PC-7 and the PC-9 is expected to allow Pilatus to build the earlier aircraft into the late 1980s.

The standard price for a PC-7 ranges between $850,000 and $900,000 with the addition of a few basic options the difference in cost. The range of options available on the PC-7, primarily avionics and communication and navigation equipment, can raise the price of the trainer by $50,000 to $150,000.

The base price of the PC-9 starts at $1.35 million and runs to $1.4 million. Because of the greater number of equipment options available to the PC-9 operator, the price could be increased by $200,000 for the trainer, Scheuber said. This higher price could include the addition of the Collins EFIS and the HUD.

Swiss certification

Since first flight of the No. 1 PC-9 prototype on May 7, 1984, Pilatus has accumulated more than 500 hr. on both aircraft. Swiss certification of the PC-9 is expected this summer. The rollout of the first production trainer is scheduled for November.

Deliveries of the PC-9 could begin late this year or early next year, although Pilatus did riot identify a customer. The company expects to reach a two monthly Production rate in 1986.

Pilatus has started production of the first batch of 10 PC-9s and also has ordered Pratt & Whitney PT6A-62 engines and components for the second production run of 10 aircraft. Scheuber said Pilatus is in discussion with all operators of the PC-7 as potential customers for the PC-9, as well as other countries not operating the earlier trainer.

PILATUS PC-9 SPECIFICATIONS

POWERPLANT
One Pratt & Whitney Canada PT6A-62 turboprop engine with a maximum rating of 1,150 shp., flat-rated to 950 shp.

PROPELLER
One Hartzell HC-D4N-2A four-blade, variable pitch with full feather, but not reversible, capabilities.

WEIGHTS	Aerobatic	Utility
Maximum ramp weight	4,982 lb. (2,260 kg.)	7,077 lb. (3,210 kg.)
Maximum takeoff weight	4,960 lb. (2,250 kg.)	7,055 lb. (3,200 kg.)
Maximum landing weight	4,960 lb. (2,250 kg.)	6,834 lb. (3,100 kg.)
Maximum zero fuel weight	4,189 lb. (1,900 kg.)	
Basic empty weight	3,715 lb. (1,685 kg.)	3,715 lb. (1,685 kg.)

DIMENSIONS
Length	33.3 ft. (10.2 meters)
Height	10.7 ft. (3.3 meters)
Wing span	33.1 ft. (10.1 meters)
Wheel track	8.3 ft. (2.5 meters)
Wing area	175.3 sq. ft. (16.3 sq. meters)
Wing aspect ratio	6.29

PERFORMANCE
Takeoff ground roll, standard conditions at 4,960 lb.	787 ft. (240 meters)
Takeoff distance to 50 ft., standard conditions at 4,960 lb.	1,412 ft. (430 meters)
Landing ground roll, standard conditions at 4,960 lb.	853 ft. (430 meters)
Landing distance from 50 ft., std. conditions at 4,960 lb.	1,625 ft. (495 meters)
Maximum cruise true airspeed at sea level	268 kt.

Maximum cruise true airspeed   300 kt.
    at 20,000 ft.
Maximum operating speed,   320 kt.
    aerobatic category
Maximum rate of climb at sea level   4,050 fpm.
Maximum range at 10,000 ft.   600 naut. mi.
Maximum range at 20,000 ft.   830 naut. mi.
Maximum operating altitude   25,000 ft.
Service ceiling   38,000 ft.

LOAD (g) LIMITS	Aerobatic	Utility
Positive	+7g	+4.5g
Negative	–3.5 g	–2.25g

All performance figures are for the PC-9 in the aerobatic configuration, unless noted. The aerobatic configuration does not include underwing stores, while the utility configuration would include underwing stores, such as extra fuel tanks.

MB-339 offers low-cost jet trainer

Robert R. Ropelewski/Venegono, Italy
January 3, 1977

Aeronautica Macchi has concentrated on low acquisition and operating costs and a simple and efficient design for its new MB-339 jet trainer, thus challenging the current philosophical trend in Western Europe and elsewhere toward high-performance, mixed-role trainer/attack aircraft.

The company dismissed a completely new design with new turbofan engine or engines early in the definition of the MB-339 after looking closely at the requirements of potential customers. It was decided instead to stick with the same basic design used in the Macchi MB-326 trainer/light attack aircraft, but with aerodynamic refinements, a somewhat larger engine and an improved cockpit.

The result, Macchi officials contend, is an aircraft with the required performance in the trainer role but with lower procurement, operating and life cycle costs than its two primary competitors, the Dassault-Breguet/Dornier Alpha Jet and the Hawker Siddeley Hawk. They further assert that the lower cost of the aircraft, coupled with its relatively high performance, make the 339 a good candidate for a trainer that smaller air forces can use for both basic and advanced flight training.

Such an application would spare air forces the cost of buying, operating and maintaining more expensive and more complicated pre-operational training aircraft. Macchi officials acknowledge that the Hawk and the Alpha Jet have the necessary performance for this advanced role, but at a higher cost than the MB-339 and with the concurrent requirement for a lower performance trainer to be used for initial flight training.

They are convinced that both the British Royal Air Force and the French air force will still need a complementary fleet of basic trainers in 'which students will log about 150 hr. of flight time before transitioning into the Hawk or the Alpha Jet.

"Alpha Jet operating costs are too [high] for the French air force to use it as a [basic and advanced] jet trainer," Giulio C. Valdonio, chief of the Macchi advanced design department, said. "This is true for the Royal Air Force with the Hawk, as well. After seeing the Alpha Jet and the Hawk flight test results, we're even more convinced now that we have made the right choice of design and engine."

This *Aviation Week & Space Technology* editor recently flew the MB-339 on two evaluation flights totaling about 2.1 hr., which included flying the aircraft from both the front and the back seats. The impression it left was that of an aircraft that is definitely easy and un-

The Aeronautica Macchi MB-339 trainer is powered by a single Rolls-Royce/Fiat Viper 632-43 turbojet engine with 4,000 lb. of static thrust.

complicated to fly, but with enough acceleration, climb and maneuvering performance to make it interesting to potential customers as a transition aircraft to higher performance operational combat aircraft.

The flights were flown from Venegono airfield in northern Italy, where the Macchi final assembly and flight test facility for, the company's MB-326 and MB-339 trainers is located. One of these flights was flown with Macchi chief test pilot Franco Bonazzi, while the other was with Macchi test pilot Riccardo Durione. A brief hop was also flown in an Italian air force Macchi MB-326 trainer to compare the performance and handling characteristics of the MB-339 with its predecessor. The difference was significant.

In general, the MB-339 is looked upon by its designers as a "second-generation jet trainer" to replace such earlier aircraft as the MB-326, the Lockheed T-33, the Fouga Magister, the British Aircraft Corp. Jet Provost and, possibly, the Cessna T-37A. Initial definition and development work was conducted in conjunction with the Italian air force, which has a requirement for about 100 aircraft to replace its older MB-326 trainers.

The replacement market for the MB-326 alone provides a strong sales potential for the MB-339. Aeronautica Macchi has sold a total of about 730 MB-326s to 12 operators in Europe, Africa, South America and Australia.

It was partly because of this success that Macchi decided to stay with the basic MB-326 fuselage and wing design, though with aerodynamic refinements that include a longer and lower nose, larger tail fin—about 25% more surface—ventral fins on the bottom rear fuselage to damp adverse yaw and dutch roll, and a new wing leading edge profile.

About nine different aircraft designs and engine configurations were studied initially, according to Valdonio. These included both single- and twin-engine versions with the 2,960-lb.-static-thrust Snecma/Turbomeca Larzac turbofan engine used on the Alpha Jet, the 5,300-lb.static-thrust Rolls-Royce/Turbomeca Adour engine used on the Hawk, the 5,200-lb.-static-thrust Rolls-Royce RB-401-06 turbofan and the 3,700-lb.-static-thrust Garrett TFE731-3 turbofan engine.

In the end, a single-engine design was selected using a Rolls-Royce/Fiat Viper 632-43 turbojet engine uprated to 4,000 lb. static thrust. The same engine is used on the single-seat attack version of the Macchi 326, the MB-326K. The design and engine selection put MB-339—maximum level speed is about 485 kt., or Mach 0.77, and maximum dive speed is about Mach 0.82. But the 339's designers considered this to be an acceptable compromise between performance and cost.

"Costs were significant," according to Valdonio. "Going from Mach 0.82 to 0.84 meant a big increase in expense—for one thing, we would have had to go to all boosted flight controls." Currently, flight controls on the MB-339 are operated manually except for servo-controlled ailerons. These can also be operated manually in case of failure of the single hydraulic system.

Although adoption of a single Adour or two Larzac engines would offer higher performance, the procurement and operating costs of an aircraft with these engines would be significantly higher than with the Viper engine, said Ermanno Bazzocchi, the company's director general-technical. In addition, Bazzocchi said, the slightly lower critical Mach number of the MB-339 allows an aerodynamically more efficient configuration of the wing in terms of aspect ratio, sweepback, thickness and wing loading.

As for the possibility of using a smaller turbofan engine, Macchi officials rejected this solution on the grounds that the benefits offered by those turbofans that would be available during the development span of the MB-339 were not significant enough to offset the higher costs of these engines.

Chief test pilot Franco Bonazzi pointed out that the specific fuel consumption of the Viper 632 engine is reasonably good—0.95-0.96 kg./hr./kg. thrust—and the Viper offers the benefit of higher thrust at high speeds and high altitudes than available turbofans. Main factor again was the cost. "Right now, the fan engines we looked at cost twice as much as the Viper," Valdonio said.

The emphasis on costs was echoed by Gianni Cattaneo, the commercial director of Aeronautica Macchi. "We're playing the drum of cost effectiveness," he told *Aviation Week & Space Technology.* "Every solution is a compromise. It may not be the golden compromise, but it is a very reasonable compromise."

Cattaneo also confirmed that the company is not pushing the MB-339 for a close air support role. The primary target of the aircraft is the trainer replacement market, which Macchi officials see as an extensive one because of the large numbers of first-generation jet trainers now reaching the end of their service life.

For this reason, simplicity is a critical factor-simplicity of operation, simplicity of maintenance and simplicity of manufacturing—to accommodate potential export customers who are likely to demand licensed production rights. Maintenance requirements for the MB-326 have had a mean value of about 3.25 manhours per hour of flight in those countries operating the aircraft, with an average availability rate of about 75%. Macchi officials expect maintainability and availability rates for the MB-339 to be about the same, if not slightly better.

Although reliability has been improved in some areas, such as avionics, since the last generation of aircraft was developed, the increased range of avionics in the MB-339 and the incorporation of the hydraulic servo-powered ailerons may offset some of the gains. Since the engine is already proved, no significant changes are anticipated in this area.

From the pilot's point of view, one important point stressed by Macchi is that the latest generation of operational combat aircraft is not as difficult to fly as previous generation aircraft, and therefore does not require the use of particularly complex trainers with difficult handling characteristics. Where possible, Macchi engineers have tried to design the MB-339 to meet U.S. Air Force Milspec F-8785B, which set strict handling standards for new aircraft.

The result is an aircraft that is easy and comfortable to operate and which has no particular limitations—as yet—for which the pilot must be on guard. The flights flown by this *Aviation Week* editor in the first MB-339 prototype—the only one to have flown so far—came after approximately 70 test flights had been logged with the aircraft by Bonazzi and Durione. The aircraft made its first flight last Aug. 12.

Prior to the *Aviation Week* flights, most of the aircraft's flight envelope had been explored. One of the remaining unfinished tasks was spin testing, so spin maneuvers were avoided during these flights.

Walkaround pre-flight inspections before the flights were routine, with checks of landing gear, flaps, flight control surfaces, engine intakes, batteries, hydraulic lines, engine exhaust and overall condition of the aircraft's skin.

Both the front and rear cockpit are comfortable and roomy, with excellent visibility forward, sideward and rearward. Nose of the MB-339 tapers downward much more than that of the MB-326, and from either the front or rear seat of the 339 the pilot cannot see the nose. Eye level for the pilot in the rear seat is about 13 in. above that of the pilot in the front seat, and the rear seat occupant has a view that is almost better than the forward pilot's.

Where the canopy rail is about even with the pilots' shoulders in the MB-326, it is several inches lower—about an inch or two above the elbows—in the MB-339, giving the pilot even more spaciousness and visibility in the cockpit.

The No. 1 MB-339 prototype is fully equipped with both UHF and VHF communications and navigation equipment, integrated attitude director indicator (ADI)/horizontal situation indicator (HSI)/flight director, transponder and low frequency/automatic direction finder radio. Control of the communication/navigation radios can be switched from one cockpit to the other by either pilot. Seats in both front and rear are Martin Baker Mk. 1T-10F ejection seats with a zero airspeed/zero altitude ejection capability.

Starting the Viper engine, which can be done only from the front cockpit, is extremely simple. With battery and master fuel switch on and the throttle set to the idle detente, it is necessary only to push the start button momentarily in front of the throttle. The engine start sequence is automatic thereafter until the engine is running at idle.

The Macchi test pilots have been using only the battery for engine starts during the flight test program, though it is possible to use an external power source instead. Battery start tests have been conducted down to temperatures as low as −20C (−4F), at which point the two batteries on the aircraft provided enough energy for four separate starts.

The one-piece canopy, which covers both the front and rear cockpit, is hinged on the right and opens by lifting up on the left side from either cockpit. A hydraulic jack on the right side between the two cockpits keeps the canopy from closing too quickly when it is pulled downward. This pilot noted a distinct difficulty in getting the canopy moving off its fully open position when trying to bring it down, and the Macchi pilots confirmed that further modifications to the jack are planned to ease the forces required here.

Running the engine up to about 60% rpm had the aircraft rolling for taxi to the runway. The first prototype has no nose wheel steering system installed yet, so this was done by using rudder pedals and brakes.

Once aligned on the runway for takeoff, a quick engine acceleration check was conducted, checking to see that the acceleration from 60 to 98% rpm took 3.5 sec. or less. Weight of the aircraft was approximately 4,200 kg. (9,060 lb.), which included about 850 kg. (1,870 lb.) of fuel. Normal maximum takeoff weight in the clean configuration is about 4,400 kg. (9,700 lb.), with a maximum usable internal capacity, including tip tanks, of 1,100 kg. (2,427 lb.).

Outside air temperature for the takeoff on both flights was about 8C (15F), and winds were negligible. With the throttle fully advanced and engine rpm at 100%, brakes were released and the aircraft accelerated quickly, lifting off from the runway after about 14 to 15 sec. of roll. The roll used about half the length of the 1,000-meter (3,300 ft.) runway at Venegono.

Aircraft was rotated at about 80 kt. and lifted off at around 110 kt. There was an initial tendency to pull the nose up too high, and. this tendency was repeated a bit later when power was advanced to start a climb after flying level for a few minutes. This was due to the fact that the mass balance of the control sticks has not yet been properly refined, and the sticks have a tendency to continue coming back when pulled aft in an acceleration and to continue going forward when pushed in that direction during deceleration. This and a number of other minor problems are being corrected while the aircraft is grounded for modifications and updating.

Climb performance

Because Venegono field is within the Milan terminal control area, there was no opportunity to conduct a maximum performance climb from takeoff. However, the climb performance of the aircraft from 2,500-30,000 ft. after leaving the control zone tended to verify the MB-339 published climb rate of sea level to 30,000 ft. in approximately 7 min., 8 min. from brake release.

On the second flight, and at maximum clean takeoff weight, the aircraft reached 40,000 ft. 15 min. after brake release. This is not an accurate measurement of performance, however, because the time included about 3 to 4 min. of level flight at 2,500 ft. while clearing the Milan control area, and the climb to 40,000 ft. was done in a slow spiral over the Alps to avoid crossing the Swiss border, which restricted the flying area on three sides.

Climb rate of 2,500 ft./min. on the vertical speed indicator when passing through 30,000 ft. was in keeping with the 2,200 ft./min. at standard conditions listed in the aircraft's technical manual. Likewise, the fuel flow meter in the cockpit read about 15-16 kg./min. at 30,000 ft., compared to listed performance of 14 kg./min. Difference was due largely to temperature.

At 40,000 ft., the nose was pushed over to a steep dive and the aircraft accelerated to its maximum speed. Mild Mach buffet began to be felt about Mach 0.78 or 0.79 and increased only slightly until the aircraft reached a maximum velocity of Mach 0.83.

Macchi pilots have noted a very slight tendency for the nose of the aircraft to tuck under, or drop, in passing through Mach 0.76, because of a shift in the center of pressure on the wings at this point. However, it was so slight as to be unrecognizable on this flight.

There is no limit speed on the use of the dive brake or speed-brake on the underside of the fuselage below the cockpit, and this was extended during the Mach 0.83 dive to slow the aircraft. Speed brake switch on the throttle must be held for a few seconds to extend the brake fully. Extension causes the nose to pitch up slightly.

Light longitudinal control forces of the stick at 40,000 ft. increased somewhat as the aircraft descended, but overall control forces, in all axes were very light at all altitudes and airspeeds. In a series of basic aerobatics beginning at around 16,000 ft., the aircraft had sufficient power and control response through all maneuvers to avoid any uncomfortable low airspeed/uncomfortable attitude situations, even for a relatively unfamiliar pilot.

Because of the light stick forces, there was a tendency to pull back a bit too much at the top and downward side of loops, bringing the aircraft into mild buffeting at the edge of an accelerated stall. Easing the stick forward slightly stopped this in each case.

Aircraft could still be flown reasonably accurately with the aileron servo-controls turned off to simulate a hydraulic failure, but in an actual situation the pilot would probably prefer to land fairly quickly because of the high stick pressures required. With the servo-control working, the aircraft had a roll rate of about 150 deg./sec. at 300 kt. and 15,000 ft., which is relatively quick for an aircraft in this category. Maximum roll rate of the aircraft is 190 deg. per sec.

Stalls in the MB-339, conducted between 10,000 and 15,000 ft., were noteworthy because of the full control of ailerons, rudder and elevator that remained. In the clean configuration, mild buffeting began at about 112 kt. and stayed at about the same level until the aircraft stalled at around 105 kt. This matched the performance charts in the technical manual for the MB-339 at that particular weight and, altitude.

In the landing configuration, with gear extended and flaps in the 62-deg. landing position, moderate buffeting began at about 96 kt. and the aircraft stalled at about 90 kt. In both clean and dirty configurations, wings stayed level during the stall as the nose pitched down about 10 deg., then entered a steady oscillation in which it went from about 10 deg. down to 5 deg. up, and then back down again.

The wing of the MB-339 has stall fences at 60% of the span on each side, and these ensure that the inner portion of the wing stalls first, allowing the pilot to retain control of the ailerons on the outer portion of the wings. It was necessary to hold the control stick in the full aft position to maintain the stalls, and the aircraft recovered from the stall as soon as the stick was released.

Aircraft stability

One other aspect of the stability of the 339 that was impressive was the almost complete absence of adverse yaw or dutch roll. The aircraft could be rolled as sharply as possible without any visible indication of adverse yaw. This is due mainly to the increased area of the tail fin on the MB-339 compared to the MB-326, which has distinct dutch roll/adverse yaw characteristics when carrying external stores. To further eliminate this problem on the 339, the new aircraft also has two ventral fins on the lower rear portion of the fuselage.

On both flights in the MB-339, the return to Venegono field was preceded by several instrument landing system (ILS) approaches to Milan's Malpensa airport. Although the 339 has no automatic pilot or other stability aids, the basic stability of the aircraft was such that aircraft could be flown smoothly and accurately under instrument flight (IFR) conditions in a fairly busy traffic pattern. The flight director eased the navigational chores considerably, although there was some chasing of the localizer beam after radar vectoring brought the aircraft across the localizer beam at too large an angle. Overall, the stability of the aircraft under these conditions, without heading hold or altitude hold or other assistance, was impressive.

From Malpensa, it was only a few minutes flight back to Venegono, where a standard break was made over the runway at 250 kt., throttle back and dropping gear and flaps during the downwind turn. Limits on extension of these items are 175 kt. for landing gear and 27 deg. of flap (takeoff position), and 150 kt. for full flap extension (65 deg.) for landing.

Initial landings at Venegono by this pilot tended to be rather cautious—somewhat fast and somewhat long. On a 3,000-ft. runway, however, these are luxuries. Bonazzi made the final landing of the first flight, flying the approach at about 105 kt. and touching down just a few feet inside the threshold at about 95 kt.

Without braking particularly hard, the aircraft was stopped in about 2,000 ft. Macchi advertises a ground roll of 1,360 ft. for the aircraft at a weight of 7,550 lb.

Brakes on the MB-339 are relatively sensitive, and without nose-wheel steering it was possible on landing to induce a swerving of the aircraft from side to side if pressure was not absolutely even on both pedals. This, however, ceases to be a problem as familiarity increases.

A number of other minor items were noted that Bonazzi and Durione acknowledged had been identified in early test flights and were being corrected. These included:

- Throttle and flap control levers—These were very stiff in both cockpits, and made fine throttle adjustments difficult. Linkages for both are being fine tuned to remove this discrepancy.
- Longitudinal trim—Thumb trim switch on the top of the control stick had to be held too long before any change in trim could be noticed. Original aim was to use the same actuators for both aileron and elevator trim. Macchi test pilots and engineers have found, however, that the elevator trim actuator needs to be about twice as fast as the aileron trim actuators, and this will be changed on future prototypes.
- Air conditioning noise in the cockpit has been identified as an annoyance factor by some pilots, and air conditioning ducts are being widened on future aircraft to reduce this noise.

Other than these items, Macchi test pilots say they have encountered no major troubles so far in the flight testing of the MB-339.

Second prototype

A second prototype, this one to pre-production standard, will join the first during the spring. Six pre-production aircraft will follow, three of these to be used in the Macchi flight test program and the other three to be used in evaluations by the Italian air force.

These aircraft will be delivered to the air force beginning around mid-1978, and will be used also, along with the first three pre-production aircraft, to define the training syllabus for the MB-339. The MB-339s will replace MB-326s and Fiat G-91s operated by the air force from late 1978 onwards.

The Australian air force, already operating MB-326s, is also considered a good potential customer for the MB-339. The Australians recently issued an operational requirement for a new trainer to enter service in 1982. About 100 aircraft are required. The Hawker Siddeley Hawk and the MB-339 are believed to be the leading contenders for this market.

Macchi officials also have begun making preliminary presentations and demonstrations of the MB-339 in the Middle East, Far East and South America.

Hawk displays
simple handling qualities

Herbert J. Coleman/Dunsfold, England
May 17, 1976

Hawker Siddeley Hawk jet trainer, now aimed at a world market for a low-cost battlefield combat fighter, is an uncomplicated aircraft that can quickly adapt its simple handling qualities into requirements for a low-altitude strike mission. The Hawk, a two-seat trainer that probably will be developed into a single seat fighter, is in advanced stages of development for the British Royal Air Force in the training role on the basis of an order for 175 aircraft.

Hawker Siddeley Aviation, however, is building a privately owned version, identical to the trainer with exception of navaids, that will fly soon and will be used for demonstrations throughout the world. The Hawk will be priced at about $2 million, and the firm is stressing to potential customers that it will strive to hold the price in that range.

The aircraft also is being pushed in world markets as a collaborative program in which Hawker Siddeley would provide management and production know-how to keep the price at the level demanded by the Royal Air Force.

In a test program primarily directed to pilot training, the Hawk has:

- Flown at Mach 1.13 in a 30-deg. dive and has successfully entered full-flap stalls at a 30-deg. angle of attack with no problems in recovery.
- Climbed to 44,000 ft. from takeoff at an average climb speed of Mach 0.7.
- Demonstrated that recovery from tight spins-Hawk has been spun up to 14 turns—can be accomplished easily by neutralizing controls and applying back pressure on the control stick.

The Hawk has completed more than half of its development test program and is being cleared for weapons and stores capability as part of its RAF operational requirement, but also for the export potential. Its Aden gun has been cleared to fire at up to 550-kt. speeds.

From a test standpoint, the most important problem is operation of the Rolls-Royce/Turbomeca Adour engine at high altitudes, above 44,000 ft., when the powerplant runs rough. The engine is being tested in a high-altitude chamber at Pyestock, England, to determine the cause of the roughness, which is not apparent at lower altitudes.

Hawker Siddeley Hawk jet trainer is fitted with Aden gun, mounted in a centerline pod, and two Engins matra rocket pods for Royal Air Force weapons tests now taking place in Britain. Ventral strake under tailplane has been enlarged to improve directional stability.

Other than the high-altitude problem, the aircraft has exceeded specifications and the operational requirement set by the RAF. Development work for the RAF will be completed by the end of this summer; about 13 Hawks will be flying by the end of October and testing by that time will be aimed at defining an operational requirement for a combat fighter. The aircraft will be built at a rate of four per month for the RAF, and export deliveries to overseas customers could begin by the end of 1977.

The Hawk is being studied closely by the U.S. Air Force after a series of test flights by USAF pilots as a possible aircraft for replacement of the Northrop T-38. An offshore buy could be a possibility under terms of the U.S.-British memorandum of understanding on sales of equipment to the Dept. of Defense.

The Hawk also has been flown by a number of Finnish air force test pilots here and the final results of a technical evaluation at both Dunsfold and Helsinki will be available by mid-summer.

Hawker Siddeley is putting top emphasis on the sale to Finland, along with keeping in close touch with the Egyptian air force on a possible buy that would involve collaborative construction. Both countries are evaluating other jet trainers, including the Franco-German Alpha Jet, on both operational and economic grounds.

Hawker Siddeley believes that the aircraft is now proved, on the basis of test flights so far, and it is centering on an economic proposal to Finland that involves:

- Introduction of the Finnish industry into world markets in Europe, North and South America and the Middle East, where the Hawker Siddeley parent group has had a long-established network of marketing and manufacturing facilities.

- Program of offset compensation to Finland that covers the sale of consumer goods, research programs and joint marketing of Finnish goods along with coproduction on the Hawk project.
- Rolls-Royce already ordering industrial paper products from Finland for use at its Derby and Bristol manufacturing centers. Those two and other major suppliers to the Hawk program— such as Lucas Aerospace, Dunlop Aviation Div., Smiths Industries and Plessey—are setting up offset agreements to promote the sale.
- Establishment of a Hawk program office in Helsinki to deal with the economic program and to act as a liaison office for the Finnish air force and various European manufacturers concerned in Hawk production.

The Finnish air force is studying Hawker Siddeley design work on transforming the Hawk into a single-seat fighter that could have a low-level battlefield capability, which could add a new dimension to the country's inventory.

Single seat configuration would involve considerable extra fuel, about 600 lb., in a bag behind the pilot to extend ferry and loiter range. Present plans also call for a variety of avionics configurations that could lead to new nose designs for the Hawk.

The company-owned civil Hawk will be certificated by the British Civil Aviation Authority and will be similar to the military Hawk, apart from navaids. The civil Hawk will be equipped with airline standard navaids since the aircraft will be used as a demonstrator for potential customers.

The Hawk will be equipped with five hardpoints for weapons de-livery, against the present three pylons required by the RAF, although that air force has retained its options to take the higher-rated version at a later date.

The Hawk was flown recently by this *Aviation Week & Space Technology* editor with A.P.S. (Andy) Jones, deputy chief test pilot, in both training and battlefield combat modes.

In the training mode, the Hawk is a simple aircraft to fly and derives much of its handling qualities from the Hawker Hunter jet fighter. Jones pointed out that the Hawk wing was developed, with inputs from the Hunter, Harrier and A300B Airbus, to a point where it is acceptable as a low-altitude fighter capable of 7-8g in tight turns.

Hawk handling is best demonstrated in spins. On this flight the spin was started at 28,000 ft. by throttling back to idle thrust, and slowing the aircraft to 180 kt. at a high angle of attack. With the stick full back, the Hawk started the spin in a flat attitude and a slow rate of turn.

After six turns, the Hawk recovered by 15,000 ft. and pullout was achieved at 5.6g. For demonstration purposes, the spin was deliberately tightened by applying opposite aileron to achieve a spin rate of 4 sec. per turn. Highest rate of spin during tests has been about 2.5 sec. per turn, although the recovery rate is less the 5 sec. once the controls have been neutralized.

Recovery from spins is positive and fast, and little rudder control is necessary once the spin has been stopped.

The Hawk can be flown supersonically, up to Mach 1.3, by putting it into a 30-deg. dive. Only control change is a slight tendency to a nose pitch-up in the transonic regime.

Stalling the Hawk is a standard procedure for students, and the aircraft resists a stall until forced into an unusual attitude. Clean stall was made by placing the Hawk in a 30-deg. angle of attack, dropping full flap and placing the stick hard back.

Even in a full stall—with noticeable buffet—the Hawk could be controlled laterally in turns to the right or left.

Stall performance is more marked at low levels, where at 5,000 ft. the Hawk can climb out of a full stall by applying full maximum power. An accelerated stall was entered at 300 kt. by holding the stick full back to the 30-deg. angle of attack and starting a tight turn, a maneuver devised, for the battlefield combat role where up to 8g turns could be an operational requirement. There is considerable buffet at low-altitude high-g turns, but the Hawk is controllable in 180-deg. turns on the edge of a stall.

The flight from Dunsfold test field was made during clear weather, but with a strong crosswind of up to 40 kt. and gusts at nearly 90 deg. to the runway. The task was to climb to 42,000 ft. at Mach 0.7 in about 10 min., and then to 45,000 ft. in 2 min.

The Hawk has been slightly modified by adding more area to its aft ventral strake, and the flight was partly to measure handling at high speeds. On one supersonic test during the flight, the Hawk was put into a 50-deg. dive at 44,000 ft., and leveled at Mach 1.05 at 36,000 ft.

There is a slight compressibility—Jones called it a cobblestone effect—as the Hawk goes through Mach 0.92, and this results in a slight loss of aileron effectiveness. However, aileron control is fully restored at Mach 0.97.

Test pilots at Dunsfold are enthusiastic about the Hawk wing design and, as Jones put it, "There is really no way to get into trouble with this airplane." The only design change in the wing has been addition of vortex generators to help increase aileron effectiveness at Mach 0.92.

Control test at 15,000 ft. for acceleration-deceleration measurement was one demonstration of Hawk stability. At 250 kt., the Hawk was taken to 460 kt., about Mach 0.85, in 96 sec. and then throttled back to 220 kt. in 156 sec. During that run, the Hawk consumed 48.5 lb. of fuel.

Hawk has excellent visibility from both front and rear seats, best shown on landing, when final approach is aimed at a 105-kt. threshold speed by allowing 1 kt. for every 220 lb. of fuel still on board.

Landing the Hawk at Dunsfold after a 1-hr. 25-min. flight, in gusty conditions involved a final approach of about 120 kt., and at flareout a gust estimated at 40 kt. hit the aircraft. There was no problem in directional control and the gust was countered by a slight drop of the wing in the direction of the wind.

Tests have shown that the Hawk wing has less induced drag than any other fighter built by Hawker Siddeley. Pilots are convinced that the wing compares in this respect with any other fighter in the world—and it now appears that the wing will establish the pattern for a family of combat aircraft.

Handling of the Hawk indicates that the aircraft will allow for simple transition in pilot training, and eventually will prove itself in a single- or two-seat combat mode in the rugged environment of the ground attack requirement.

Test pilot Jim Hawkins, who is responsible for weapons testing, said an outboard wing fence and the vortex generators solved a problem of instability and the end result was that the Hawk is now a high- and low-altitude weapons platform that will pose few problems for pilots new to the aircraft. He also said that the Hawk has been spun with weapons stores attached to wing points, and the installed tail chute was never needed.

The only engine problem is loss of thrust at high altitude, now under investigation, and this is a safety concern only if the Hawk is put into a high angle of attack above 34,000 ft. Relights have been made at high altitudes, above 25,000 ft., using the Micro-Turbo gas generator starter.

Both Jones and Hawkins said that the Hawk trainer program has derived large benefits from operational experience of the Franco-British Jaguar strike aircraft, also powered by the Adour engine. The Adour for Hawk is now cleared for 300 hr. between overhauls and will be advanced to 400 hr. by the time it enters RAF service next year.

Navy T-34C trainer in production

Donald E. Fink/Wichita
April 12, 1976

Beech Aircraft Corp's new T-34 Turbo Mentor Navy trainer, which matches the Pratt & Whitney PT6A-25 turboprop engine with the airframe from Beech's earlier piston-powered T-34B, has both the solid handling characteristics of its predecessor and the performance of an advanced transition trainer.

The Navy is organizing a new integrated flight training syllabus to exploit the capabilities of the T-34C by combining the primary and basic phases in the present flight training program into a single basic phase. In its dual role as the Navy's new basic trainer, the T-34C will replace both the older T-34Bs and the Navy/Rockwell T-28As. Production versions of the T-34C the first of which is scheduled to fly next June, will have new airframes built to the Navy's latest fatigue lifetime specifications. But except for the turboprop powerplant and minor airframe changes associated with the new engine installation. the T-34C looks very similar to the earlier B model.

The same basic configuration was retained for the C model because Beech designers found that even if they had started with a clean sheet of paper they would have ended up with a design that looked like the T-34, according to Robert R. Stone, senior engineering test pilot at Beech.

Separate canopies are provided in the tandem cockpit for the instructor (rear) and the student pilot. Note the fuselage strakes located aft of the rear cockpit and the ventral fins, which were added in order to improve spin recovery characteristics of the longer T-34C airframe.

Retaining the basic T-34 airframe enabled Beech to eliminate considerable initial research and development work that would have been required to prove an all-new design, thereby keeping the program within the rather severe budgetary limitations imposed by the Navy. The company also was able to develop two inexpensive YT-34C prototypes by modifying T-34B airframes provided by the Navy.

The performance of the new trainer configuration was evaluated here recently by this *Aviation Week & Space Technology* editor flying one of the YT-34C prototypes with Stone in the rear seat. The evaluation flight was planned in the No. 2 prototype, which is closest to the production configuration, but a last minute mechanical problem grounded the aircraft.

The evaluation flight was made in the No. 1 prototype, which is still fitted with much of the instrumentation used in the initial flight test program. Aside from some performance deficiencies caused by the test installations and some differences in cockpit controls, the No. 1 prototype is a fairly good representation of what the production T-34C will be, Stone said.

The YT-34C airframe is slightly larger than the T-34B in all three major dimensions as a result of the increased length of the PT6A-25 powerplant, the slightly longer span of the Beech Baron wing, which is used in place of the Bonanza wing on the earlier model, and the additional vertical tail area required.

Overall length of the C model is 28.7 ft., the wingspan is 33.3 ft. and the height at the tail is 9.8 ft. Aircraft empty weight is 2,940 lb. and the maximum gross takeoff weight in the Navy configuration is 4,300 lb. The airframe is stressed for +6g and −3g maneuvers.

The PT6A-25 engine was chosen for the C model because of its established reliability record and its long lifetime characteristics, Stone said. The engine also has a dry weight of 318 or 331 lb.—depending on whether magnesium or aluminum castings are used—which is slightly less than that of the 225 hp. Teledyne Continental 0-470-13 reciprocating engine used in the T-34B.

In the T-34C, the 550 shp. PT6A-25 is operated at a derated 400 shp. to enhance its fuel economy and extend its time between overhaul to at least 5,000 hr., according to Stone. The 400 shp. rating is achieved at 955 ft./lb. of torque and 2,200 rpm, so a torque limiter is included ill the engine control loop to restrict it to a maximum of 1,000 ft./lb.

At the 4,300-lb. gross weight, 400 shp. does not give the T-34C a high power-to-weight ratio, but it is sufficient for the basic training mission defined by the Navy, he said.

Beech also is offering an export version of the T-34C as a combination trainer/light ground attack aircraft. As an armament systems trainer, the T-34C will be certified at a maximum gross takeoff weight of 5,400 lb. with slightly reduced maneuvering g-force limits. It has four wing stores stations with a maximum capacity of 1,200 lb. and is offered with a PT6A-25 engine flat rated at 550 shp.

Following a walk-around inspection and a short cockpit briefing, the engine start procedure was initiated by ensuring that the power lever was in the idle position, the propeller in feather and the fuel lever in the shut-off position. In the production configuration, the fuel control function is incorporated in the propeller lever and the fuel is shut off by pulling the lever back through the feather position.

With brakes locked, the battery switch was turned on and the starter/igniter button pushed until the gas generator speed N_1 reached 12%. The fuel lever then was advanced to the full on position-in production versions this function will be accomplished by advancing the propeller lever to full increase- and the starter/igniter was held until the intermediate turbine temperature (ITT) peaked and returned to the normal operating range.

Temperature peaked at about 740C and then dropped to about 600C, which is within the normal operating arc. If the temperature appears headed for a peak near the 1,200C maximum on the ITT gauge, indicating a hot start, the fuel is shut off and the starter/igniter is held in the engaged position until the temperature drops.

Stone said the requirement to watch for a temperature peak is good training. for students who will be transitioning into jets or advanced turboprop-powered aircraft. He added that he had not encountered a hot start in several years of flight tests with the YT-34C prototypes.

Following the ITT peak, the starter/igniter switch was released and the starter automatically switched to the generator mode. An automatic starter/igniter switch is engaged whenever the T-34C is to be flown aerobatically.

Once the engine was started, the propeller lever was moved to full increase, and the engine was controlled only with the power lever. This also will help develop habit patterns the student pilot will use when flying pure jets. The engine is canted to the right and downward to offset the torque roll tendency, Stone said. The resulting lack of directional effects simulates the centerline thrust of jet engine installation. The three-bladed Hartzell hydraulic propeller on the T-34C has the full beta range, and reverse pitch can be used to assist braking during taxi or landing rollouts.

The Navy excluded nosewheel steering from the T-34C design as another austerity measure. Despite this, the YT-34C demonstrated good ground handling characteristics using rudder and differential braking. Taxi-back following the final landing was hampered by loss of left brake pedal effectiveness in the front cockpit, but Stone attributed this to a mechanical problem unique to the No. 1 prototype.

Following pretakeoff checks, the YT34C was taxied onto Runway 18 at Beech's flight test facility. The Wichita area was experiencing heavy thunderstorm activity at the time of the flight, and variable gusting wind conditions prevailed on the runway.

Takeoff power was applied smoothly, both to prevent surging in the propeller governor and to ensure that engine torque limits were not exceeded. The No. 1 prototype is not fitted with an automatic torque limiter.

The nose was rotated at 80 kt. and the aircraft flew off the runway almost immediately. Initial acceleration was quite fast and the best rate-of-climb airspeed of 120 kt. was reached by the time the gear was retracted. The vertical speed indicator showed an average rate-of-climb of about 1,300 fpm during a climb to an altitude of 2,500 ft.

The demonstration flight was limited to a maximum of about 3,000 ft. by a heavy overcast layer, but this was sufficient to evaluate the basic handling characteristics of the aircraft. Roll response of the YT-34C is quite positive, due in part to the new aerodynamically balanced ailerons that assist lateral control inputs.

The aircraft also has docile stall characteristics, with the break occurring at about 73 kt. in a wheels up/flaps up configuration and idle power. In power stalls with the gear and flaps up the break occurs at about 59 kt. Stalls with gear and flaps extended occur at about 60 kt. at idle power and 51 kt. with power.

Since this pilot is not aerobatic qualified, Stone demonstrated the capabilities of the YT-34C with an abbreviated version of the routine he flew during the 1975 Paris air show. This included a series of right and left turn rolls and vertical zooms which terminated in hammerhead stalls and vertical dives to recover airspeed. The PT6A engine ran smoothly throughout the vertical maneuvers. The initial sequence was followed by a loop with a snap roll on top and a vertical entry into an inertial coupled three-turn spin.

An extensive series of spin performance evaluations had to be flown early in the prototype flight test program because wind tunnel tests had shown the lengthened YT-34C fuselage had substantially different spin characteristics from that of the T-34B. The increased propeller disk area and the increased fuselage area forward of the center of pressure made it necessary to add strakes to the aft fuselage, extending forward from each horizontal stabilizer, and small ventral fins on each edge of the aft fuselage.

Following a final series of rolls, Stone stopped the aircraft in the inverted position and cruised in negative g flight for about 45 sec. As he rolled the aircraft upright, the engine surged, indicating the inverted engine protection system had been triggered by the centrifugal force of the roll.

The PT6A-25 engine has modified fuel and oil systems which permit up to 2 min. of inverted flight. When fuel in the inverted sump tank is exhausted—after about 2 min.—the engine flames out and the fuel control automatically switches to idle. At the same time, the starter/igniter comes on and provides spark for a relight.

About 30 sec. after engine flameout, the inverted oil sump drains and the spring-loaded propeller control mechanism moves the blades to the feathered position. All the pilot has to do to recover is roll upright and the engine automatically reignites, Stone said. Once oil pressure returns to the propeller hub, the blades move to the full increase position.

Production instrument panel on the Navy/Beech T-34C, shown above in the electrical system mockup, has instrument groupings patterned after those of advanced jet fighters in order to ease the transition of student pilots to higher performance aircraft. Note the power lever and combined propeller/fuel level on the left.

Navy/Beech T-34C prototypes are modified T-34Bs fitted with Pratt & Whitney PT6A-25 turboprop engines. No. 2 T-34C prototype (foreground) demonstrates the aircraft's inverted flight capability.

If the aircraft is rolled upright within about 10 sec. after engine flameout, the gas generator usually will relight on internal heat, he said. If the engine is allowed to cool, the automatic igniter, which can be selected in any flight altitude when the engine drops below 400 ft./lb. of torque, will relight the gas generator.

After the aerobatic demonstration, this pilot flew back to Beech field for an evaluation of the YT-34C in the traffic pattern. Two approaches to touch and go landings were required to get the landing sequence within acceptable performance limits. An additional series of touch and go landings was made with a steadily increasing proficiency in pattern flying and landing accuracy. The YT-34C is equipped with an angle-of-attack indicator set for a nominal approach speed of 80 kt. The angle-of-attack sensor is mounted on the left wing leading edge outside the propeller slipstream. At low landing weights, the approach can be flown at 76 kt.

In Navy-style high-sink-rate landings, the 80 kt. is held to touchdown. The airframe and landing gear on production T-34Cs is designed to withstand landing sink rates of 13.8 fps. For conventional landings, the power lever is eased off prior to the final flare and the aircraft touches down smoothly at about 60 kt.

Reverse pitch was used to reduce the final landing rollout and the aircraft was taxied back to the parking ramp and shut down.

Student pilots starting their training in the T-34C should find it a relatively simple aircraft to handle in the type of basic maneuvers that will be taught in the initial phases. As they progress to the advanced phases, the higher performance characteristics of the aircraft, including a projected maximum cruise speed of 226 kt. at 17,500 ft. and the excellent acrobatic capabilities, should ease their transition into larger aircraft.

In the Navy's present training program, all student pilots are given 26 hr. of primary instruction in T-34Bs before they are sent to specialized training in jets, multi-engine aircraft or helicopters. Jet students get 118 hr. of basic flight training in the Navy/Rockwell T-2 Buckeye and 116 hr. of advanced training in the Navy/McDonnell Douglas TA-4J.

Multi-engine students are given 90 hr. of basic instruction in the Navy/Rockwell T-28A and 105 hr. of advanced training in the Navy/Grumman TS-2A trainer version of the S-2A aircraft. Helicopter students are given 90 hr. in the T-28A and then 30 hr. in the Navy/Bell TH-57A and 65 hr. in the Navy/Bell TH-1L.

Under the new integrated flight test syllabus, all student pilots will fly 65 hr. in the T-34C. Pilots going into the jet category, which will be redesignated strike training, will be given 100 hr. of intermediate training in the T-2 and 90 hr. of strike phase training in the TA-4J. Maritime (formerly the multi-engine category) and helicopter pilots will be given an additional 26 hr. in the T-34C before progressing into the advanced aircraft in their categories.

A total military market for about 400 aircraft is forecast for the T-34C, including possible applications in the USAF inventory. Export sales potential of the aircraft could equal that of military sales.

Four production flight test aircraft are in various stages of fabrication and early assembly at Beech. The first two will be flight test aircraft and the second two will be used for reliability and maintenance demonstrations as well as flight tests. Static and fatigue test articles also are being built. No. 5 aircraft will be the first production delivery model.

The Navy has 93 T-34Cs on firm order from Beech and has an additional 23 on optional order for the Fiscal 1977 transition period. Orders for a total of 109 aircraft had been scheduled for Fiscal 1977, but this may be cut to 33. Fiscal 1978 procurement may then be increased to 103, depending on the availability of funding.

Budgetary restrictions on the T-34C program forced Beech to pursue a unique procurement approach on the avionics suit for the aircraft in an effort to keep the purchase price as low as possible. Navy said it wanted tactical air navigation (Tacan) and UHF communications, with backup for both units, in the T-34C. But Navy said it could not afford expensive military specification systems.

Several suppliers of general aviation avionics systems offered to develop inexpensive Tacan units by modifying commercial omni distance measuring equipment (DME) units to also display bearing information. Beech finally chose an integrated package offered by Rockwell International's Collins Radio Group, which includes the DME/Tacan, an inexpensive UHF radio with a split turning head for the front and rear cockpits, and an omni receiver. The omni receiver functions both as a backup for the Tacan and as a backup for the UHF communications system. Collins also is supplying dual transponder units for the aircraft.

Provision has been made in the system for area navigation (R-Nav), which Navy may add to the aircraft next year if funding is available. There are only two Tacan stations in the vicinity of Navy's Flight Training Center at Pensacola, Fla., and these often are used by up to 30 student pilots at a time.

With R-Nav, each instructor/student team can be given different bearing/distance offset coordinates that will enable it to use the system in relatively traffic-free airspace.

Using the integrated avionics package from Collins has enabled Beech to develop a full wiring harness for the T-34C on a separate mockup board and to resolve fit problems with a fuselage mockup that has a production standard cockpit. Once the T-34C is in production, the wiring harnesses will be assembled on the separate boards, back-wired with instrument connector plugs and given a full series of computer-operated checkout tests.

The wiring harness is designed for installation by one person on the final assembly line in about 30 min. If an electrical problem develops in the field, the harness can be removed easily and replaced by a new one, limiting the maintenance down time on the aircraft.

The same characteristics that made the T-34C attractive to the Navy—high performance coupled with low initial unit cost and reduced operating and maintenance costs—also are attracting the interest of foreign air forces, according to R.E. Staggs, manager of international project sales at Beech.

Morocco has ordered 12 aircraft and negotiations for additional orders are under way in several other countries. Beech has 10-15 delivery positions open in 1977—the production rate is expected to reach six a month next year—but large numbers would not be available to international customers until 1978.

Beech's regional sales managers have been offering the T-34C as part of the company's product line for the past year or so, Staggs said. The international T-34C sales effort will be increased this year as the first production models become available for demonstration purposes. Initial deliveries to Navy will start early next year.

South America is considered an excellent market for the T-34C because of the various air force modernization programs under way there, he said. Beech sold a number of T-34As, marketed internationally as the B-45, to several South American countries when their air forces were flying the North American P-51 World War 2 fighter and the F-86 jet.

These countries now are modernizing their air force inventories with the USAF/Northrop F-5 and the Dassault Breguet Mirage series fighters and are in need of a high-performance transition trainer, Staggs said.

Beech also sold the B-45 to several countries in the Far East, including Japan which produced the aircraft under license from 1958 through 1962, and follow-on orders in that area of the world for a T-34C version are considered possible.

A wide range of armament systems is being offered on the export versions. The four hard points, which Beech engineered into the basic T-34C even though Navy did not state a requirement for them, are designed for an external load of 1,200 lb. but versions with external load capacities of 2,000 lb. and more are being offered to some international customers.

On most export versions, the engine torque controller will be modified to make the full 550-shp. potential of the PT6A-25 powerplant available.

Index

Illustrations are in **boldface.**

Other Bestsellers of Related Interest

Business & General Aviation Aircraft Pilot Reports
Aviation Week & Space Technology Magazine
Read all about the most interesting general aviation aircraft today from these first-hand accounts from America's premier aviation writers.
0-07-003092-8 $19.95

Commercial & Regional Transport Aircraft Pilot Reports
Aviation Week & Space Technology Magazine
Read all about the most interesting air carrier aircraft today from these first-hand accounts from America's premiere aviation writers.
0-07-003167-3 $21.95

Helicopter Pilot Reports
Aviation Week & Space Technology Magazine
Read all about the most interesting helicopters from these first-hand accounts from America's premier aviation writers.
0-07-003168-1 $21.95

How to Order

Call 1-800-822-8158
24 hours a day,
7 days a week
in U.S. and Canada

Mail this coupon to:
McGraw-Hill, Inc.
P.O. Box 182067
Columbus, OH 43218-2607

Fax your order to:
614-759-3644

EMAIL
70007.1531@COMPUSERVE.COM
COMPUSERVE: GO MH

Shipping and Handling Charges

Order Amount	Within U.S.	Outside U.S.
Less than $15	$3.50	$5.50
$15.00 - $24.99	$4.00	$6.00
$25.00 - $49.99	$5.00	$7.00
$50.00 - $74.49	$6.00	$8.00
$75.00 - and up	$7.00	$9.00

EASY ORDER FORM—
SATISFACTION GUARANTEED

Ship to:

Name _____

Address _____

City/State/Zip _____

Daytime Telephone No. _____

Thank you for your order!

ITEM NO.	QUANTITY	AMT.

Method of Payment:

☐ Check or money order
enclosed (payable to
McGraw-Hill)

☐ VISA ☐ DISCOVER

☐ AMERICAN EXPRESS Cards ☐ MasterCard

	Shipping & Handling charge from chart below
	Subtotal
Please add applicable state & local sales tax	
	TOTAL

Account No. ☐☐☐☐☐☐☐☐☐☐☐☐☐☐☐☐☐

Signature _____ Exp. Date _____
Order invalid without signature

**In a hurry? Call 1-800-822-8158 anytime,
day or night, or visit your local bookstore.**

Code = BC15ZZA